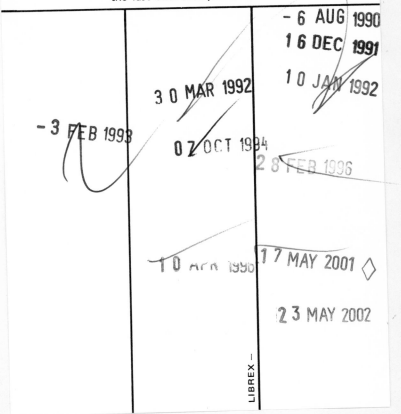

Petroanalysis '87

Petroanalysis '87

Developments in
Analytical Chemistry
in the
Petroleum Industry

Edited by
G. B. Crump

John Wiley & Sons
Chichester · New York · Brisbane · Toronto · Singapore

Copyright © 1988 by John Wiley & Sons Ltd.

All rights reserved.

No part of this book may be reproduced by any means, or transmitted, or translated into a machine language without the written permission of the publisher

British Library Cataloguing in Publication Data available:

ISBN 0 471 91946 2

Printed in Great Britain.

Contents

List of Contributors vii

Preface ix

Future Trends in Petroanalysis
L.S. Bark 1

Laboratory Integration: Automation and Robotics
G.B. Wilson 11

Discharge - Thermospray Ionisation of Lubricant Components
M. Thompson 19

Advances in Mass and N.M.R Spectroscopies to Characterize Heavy Petroleum Fractions
M. Bouquet, D. Jolly, A. Bailleul and J. Brument 33

The Analysis of Volatile Organic Compounds using Thermal Desorption/GLC/MS
P.J.C. Tibbetts, A.J. Holland and R. Large 37

Improvements in Oil Fingerprinting: GC.HRMS of Sulphur Heterocycles
P.J.C. Tibbetts and R. Large 45

Inductively Coupled Plasma/Mass Spectrometry
A.A. van Heuzen 59

Geochemistry - More Analyses Give a Better Picture
J.R. Gray 69

Environmental Monitoring of Trace Elements in Water Discharges from Oil Production Platforms
C.B. McCourt and D.M. Peers 77

FABMS Analysis of Surfactants and Polar Petroleum Compounds
R. Large, P.J.C. Tibbetts and A.J. Holland 99

POSSUM - A Method for the Determination of Light Ends in Unstabilised Crude Oils and Condensates
H. Fitzgerald 113

Analytical Developments in the Monitoring of Atmospheres Associated with Oil Based Drilling Fluids
J.W. Hamlin, M.H. Henderson and K.J. Saunders 125

Automated Identification of Petroleum Refinery Streams
D.J. Abbott 135

Oil Analysis and Machine Condition Monitoring Techniques
R.F.W. Cutler 139

Wax Chromatography - The 80's Crossroads
A. Barker 159

Thermal Analysis - I.P. Studies
S.R. Wallis 173

The Use of Inductively Coupled Plasma Emission Spectroscopy (ICES) in Residual and Distillate Fuel Analysis
B. Pahlavanpour and E.K. Johnson 189

HPLC Determination of Aromatics in Middle/Heavy Distillates A Computerized Approach
P. Richards 211

Ion-Chromatography in the Oil Industry
D. Mealor 221

The Study of Lubricating Oils and Additives by Freeze-Fracture Replication Transmission Electron Microscopy - FFRTEM
K. Reading, A. Dilks and S.C. Graham 239

The Analysis of Polychlorinated Biphenyls (PCBs) in Mineral Based Insulating Oils by Capillary GLC
K.J. Douglas and R.A. Pizzey 253

The Application of Chemometrics to ^{13}C NMR Spectra of Hydrocarbon Mixtures
A.G. King and J.M. Deane 261

Tough Problems - Novel Solutions (Abstract only)
D. Betteridge 279

Developments in the Analysis of Petroleum Products by Capillary Chromatography (Abstract only)
C.A. Cramers 281

Characterization of Heavy Oil Residues by Multi-Element Evolved Gas Analysis (Abstract only)
A.J. Meruma, M.C. van Grondelle, H.C.E. van Leuven and L.L. de Vos 283

Development of an HPLC Method for the Determination of Nitrogen Containing Corrosion Inhibitors in a Mixed Hydrocarbon/Glycol Matrix (Abstract only)
E.H. McKerrell and A. Lynes 285

Index 287

List of Contributors

ABBOTT, D.J., *Esso Research Centre, Milton Hill, Abingdon, Oxon OX13 6AE*
BAILLEUL, A., *Total-France, B.P. 27, 76700 Harfleur, France*
BARK, L.S., *Department of Analytical Chemistry, University of Salford, Manchester, M5 4WT*
BARKER, A., *Dussek Campbell Limited*
BETTERIDGE, D., *BP Research Centre, Sunbury-on-Thames, Middlesex, TW14 7LN*
BOUQUET, M., *Total-France, B.P. 27, 76700 Harfleur, France*
BRUMENT, J., *Total-France, B.P. 27, 76700 Harfleur, France*
CRAMERS, C.A., *Laboratory of Instrumental Analysis, Eindhoven University of Technology, POB 513, 5600 MB Eindhoven, The Netherlands*
CRUMP, G.B., *Secretary, Petro Analysis '87 Trust, Fernie, Chapel Lane, Willington, Nr Tarporley, Cheshire CW6 0PH*
CUTLER, R.F.W., *Wearcheck Laboratories, Robertson Research International, Llandudno, Gwynedd, LL30 1SA*
DEANE, J.M., *AFRC Institute of Food Research, Bristol*
DE VOS, L.L., *Koninklijke/Shell-Laboratorium, Amsterdam, (Shell Research BV), Badhuisweg 3, 1031 CM Amsterdam, The Netherlands*
DILKS, A., *Shell Research Ltd., Thornton Research Centre, PO Box 1, Chester CH1 3SH*
DOUGLAS, K.J., *Castrol Research Laboratories, Pangbourne, Reading RG8 7QR*
FITZGERALD, H., *Moore, Barrett & Redwood Ltd, Rosscliffe Road, Ellesmere Port, Cheshire*
GRAHAM, S.C., *Shell Research Ltd., Thornton Research Centre, PO Box 1, Chester CH1 3SH*
GRAY, J.R., *B.P. Research Centre, Chertsey Road, Sunbury-on-Thames, Middlesex TW16 7LN*
HAMLIN, J.W., *BP Chemicals Limited, Belgrave House, London SW1W 0SU*
HENDERSON, M.H., *B.P. Research Centre, Chertsey Road, Sunbury-on-Thames, Middlesex TW16 7LN*
HOLLAND, A.J., *M-Scan Limited, Silwood Park, Sunninghill, Ascot, Berks SL5 7PZ*

JOHNSON, E.K., *77 Stockbridge Road, Winchester, Hampshire, SO22 6RP*
JOLLY, D., *Total-France, B.P. 27, 76700 Harfleur, France*
KING, A.G., *Esso Research Centre, Milton Hill, Abingdon, Oxon OX13 6AE*
LARGE, R., *M-Scan Limited, Silwood Park, Sunninghill, Ascot, Berks SL5 7PZ*
LYNES, A., *Shell Research Centre, Thornton Research Centre, P.O. Box 1, Chester CH1 3SH*
MCCOURT, C.B., *Shell Research Centre, Thornton Research Centre, P.O. Box 1, Chester CH1 3SH*
MCKERRELL, E.H., *Shell Research Centre, Thornton Research Centre, P.O. Box 1, Chester CH1 3SH*
MEALOR, D., *B.P. Research Centre, Chertsey Road, Sunbury-on-Thames, Middlesex TW16 7LN*
MERUMA, A.J., *Koninklijke/Shell-Laboratorium, Amsterdam, (Shell Research BV), Badhuisweg 3, 1031 CM Amsterdam, The Netherlands*
PAHLAVANPOUR, B., *Caleb Brett Laboratories, Kingston Road, Leatherhead KT22 7LZ*
PEERS, D.M., *Shell Research Centre, Thornton Research Centre, P.O. Box 1, Chester CH1 3SH*
PIZZEY, R.A., *Castrol Research Laboratories, Pangbourne, Reading RG8 7QR*
READING, K., *Shell Research Centre, Thornton Research Centre, P.O. Box 1, Chester CH1 3SH*
RICHARDS, P., *Carless Solvents Ltd, Harwich Refinery, Refinery Road, Parkeston, Harwich, Essex, CO12 4SS*
SAUNDERS, K.J., *B.P. Research Centre, Chertsey Road, Sunbury-on-Thames, Middlesex TW16 7LN*
THOMPSON, M., *Shell Research Centre, Thornton Research Centre, P.O. Box 1, Chester CH1 3SH*
TIBBETTS, P.J.C., *M-Scan Limited, Silwood Park, Sunninghill, Ascot, Berks SL5 7PZ*
VAN GRONDELLE, M.C., *Koninklijke/Shell-Laboratorium, Amsterdam, (Shell Research BV), Badhuisweg 3, 1031 CM Amsterdam, The Netherlands*
VAN HEUZEN, A.A., *Koninklijke/Shell-Laboratorium, Amsterdam, (Shell Research BV), Badhuisweg 3, 1031 CM Amsterdam, The Netherlands*
VAN LEUVEN, H.C.E., *Koninklijke/Shell-Laboratorium, Amsterdam, (Shell Research BV), Badhuisweg 3, 1031 CM Amsterdam, The Netherlands*
WALLIS, S.R., *Castrol Research Laboratories, Pangbourne, Reading RG8 7QR*
WILSON, G.B., *Esso Petroleum Co Ltd, Esso Research Centre, Nr Milton Hill, Abingdon, Oxon OX13 6AE*

Preface

G. B. Crump – *Chairman IP Standardisation Committee*

This book contains the Proceedings of the Third Petro Analysis Symposium, held under the auspices of the Institute of Petroleum, London and the NW Region Analytical Division of the Royal Society of Chemistry.

The first PA Symposium was held in 1973, the second was in 1980 and here we are at number three. The seven year span between meetings is rather accidental. The spaces were chosen to allow for changes in analytical techniques and practices to become established, or to make significant headway, and also to hold meetings to assess the impact of these changes on the petroleum industry. A five year span was first envisaged, but I feel that seven years (the analytical itch as it were) is needed to see the maximum effect of major changes.

The scene behind PA symposia has been one of complex and dramatic changes. Through the 70's the double hoick in oil prices brought much of the industrialised world to its knees. The recycling of petroleum revenues boosted the economies of underdeveloped countries, but at a price the Banking Community (and hence the rest of us) might yet regret. A year or so ago, and through much of the past year, the oil price plummeted from heights of $30 a barrel to $9 a barrel, as the beleaguered 'rich nations' fought back against the OPEC ransom, substitution, economy, and new sources were the order of the day. As of this week we stand on a plateau of $18–$20, but for how long?

The above brief history indicates how the world continues to be hooked on oil as its main energy source, and the base for crucial products such as fuels, lubricants and petrochemicals.

With the heightening price came increased demands for quality. Moreover the winning, shipping and use of petroleum and its products has led to significant environmental problems of which you are all aware. The flexibility of the petroleum industry e.g. its ability to produce cheap leadless gasoline, and low sulphur diesel fuels and HFO's, has been a major strength. The core business has prospered—transportation fuels, lubricants and bitumen. Alongside this another hydrocarbon source, natural gas, has expanded as the UK's (and other countries) main heating source.

In all the above there is the clear connotation, that the industry serves a myriad of customers, who have become ever and ever more quality conscious. The role of the Institute of Petroleum has been to monitor, report on, and guide the changes in the industry (be it exploration or refining or marketing)—and to be a forum for debate on the above, such that the public and government agencies can have a substantially unbiased view of what the Industry is about in the UK. Many of you here are initiated in this arcane subject—but believe me, although Mr Public drives a motor car, flies in planes, wears synthetic clothes and uses polymeric kitchen utensils, etc., he is unlikely to know a carbon – carbon double bond from a marriage bond.

Nevertheless, the industry seeks to provide him with high grade products which will serve his purposes well without damage to utilities—all these assume correct usage.

In support of the above it is necessary to supply standardised test methods to validate product quality. This is the IP's strength, for it has a suite of some 350 methods, tailored to this end. Methods are not given to us on tablets of stone, they must be evolved step wise from basic ideas, with much work in committees of ST, and much work in laboratories of industry and the customer to produce correlation data. A method without precision data is not a valid method for correlative use (e.g. in product litigation)!

The world moves on and methods that were good five years ago are often redundant now. Colorimetric methods for phenolic anti-oxidants in kerosine, excellent in their day, have been replaced by HPLC, as has measurement of total aromatics in middle distillate fuels and lub oils.

Clearly a focus of this meeting is to review new developments of technique—leading to new methods. It also covers the re-appraisal of existing methods for more meaningful data, and the search for methods leading to more economic provision of data—and even, horror of horrors, deskilling complex operations, so that there is no human analytical intervention between the sample and the production of test data on it! An Orwellian concept perhaps.

This book will show how far we have gone along this road—and also provide an updated view of other difficult problems besetting the industry—where the petroleum analyst is still assuredly the only person who can solve them.

Future Trends in Petroanalysis

L.S. Bark
Department of Analytical Chemistry, University of Salford, Manchester, M5 4WT

There are always several problems associated with the preparation of a lecture such as this; the main one is to design the lecture in such way as to avoid the audience being able to answer the question "Is this a fool or a brave man, who attempts to postulate the future?"

I realised that part of the question was answered when I agreed to give the lecture and I hope that I will be brave enough to be able to continue to the end and be fortunate enough to be about 5-10% correct in my postulations.

Before grasping the nettle of the future trends I feel that it is necessary to attempt to answer one of the oldest questions posed to a practising analyst:- "Why are the analyses made?"

The crude, generally chemical, analyses of the earliest days were the explorations of an unknown world, the charting of naturally occurring materials and of the relatively simple compounds that were made by man. As Science progressed, analysts produced the results upon which the information for the establishment of the disciplines of the sciences associated with the petroleum industry, viz:- chemistry, geology and other such disciplines, are structured. Techniques for the analysis of materials have rapidly increased in number, complexity and cost and we have generated methods which produce results rapidly in order to meet the ever increasing clamour for the speedy production of the information necessary to control the industry.

Let us accept the purpose of analysis, especially that for industry and for the purposes of this lecture that associated with the petroleum industry, is essentially to help in the task of making the "bottom line" of the financial proceedings "totally acceptable" or at the worst "more acceptable" to the industry.

The bottom line will take into account not only the costs of exploration, production, and disposal through sales etc, but also the costs associated with the cleanup processes necessary for the making of environmental peace and the cost of obtaining good public relations, freedom from litigation and many other non chemical items which are part of the modern petroleum industry.

Having indicated why the analysis are being done, it is necessary to postulate what are the analyses that will be required in the future. To do this it is necessary to assess what the probable areas of development in the petroleum and allied industries.

There are several areas that spring to mind:-

1. There will continue to be problems associated with exploration and the analysis of the crude oils that are discovered and used.

2. There is a growing interest in the production of oils and related products from shales, from coals and from those parts of the oil residues that are at the present not always efficiently used as feedstock for the more commercially desirable fractions. This area of production will provide many problems for the analyst.

3. The severe problems encountered with enhanced oil recovery from already tapped sources will require further investigation and research.

4. The problems associated with corrosion and linked phenomena, need attacking on a long-term basis, the aim must be prevention rather than removal of the corrosion, if costs are to be lessened.

5. The production of secondary materials from oils and from petroleum, i.e. petrochemicals, will continue to form a major area of research and development.

6. As production increases and public awareness of ecological resources develops, problems arising from environmental considerations will require new thinking, new methods and new solutions. Probable litigation will require greater accuracy in establishing sources and amounts of contaminations, as well as better monitoring of the preventative measures that will probably be enforced.

There are, and will be others but it is not possible to deal in a short time with all of the problems which have been envisaged and I intend to restrict my remarks to consideration of some of the above areas.

Obviously associated with all analysis is the question of the validity and acceptance of the results both by the analysts and by the clients so that the results can be readily and honestly be used in the solution of the equation to the "bottom line".

This concern with the validity and the use of the results brings to notice the associated question of:-

"What is actually being analysed and how meaningful is the information that has been derived from the results?"

It poses to the analyst three main questions for the future:-

1. Can methods be devised which can provide the analysis of things as they are i.e. AS IS ANALYSIS?

That is, can we analyse systems, individual compounds and materials as they are and not required to break them down into manageable pieces or transform them by chemical reaction into substances, which we are able to analyse and then postulate from the results obtained on these "alien" species, solutions to problems associated with the original species and systems.

Most systems exist in a state of dynamic equilibrium and whilst the rate of change of some systems is low, removal of one component from the system at a rate of the same order as that of the change of equilibrium cannot give either a representative sample or help to provide an answer which truly reflects the real system.

2. Can these methods be devised BEFORE the information is required and so allow the research to be problem orientated rather than be problem driven?

3. Can we obtain an intellectual interface between the analytical chemists and the consultants and the producers and users, so that the problems of each groups are known to and understood by the other group?

In some instances there will have to be a very well structured educational programme so that there is no barrier imposed by an apparently common language.

EXPLORATION AND GEOCHEMICAL ANALYSIS

There are aspects of this area of analysis, which are in need of development. It should be remembered that whereas, good work by the petroleum analyst may result in more successful economic use of a natural resource, good work by the petroleum geochemist will hopefully increase availability of natural resources.

For each of them, as for any other type of analyst, there is a similar question:-

Does the sample provided truly represent the problem to be solved?

There are significant differences between the two areas of analysis and these may be attributed to:-

1. The amount of analyte in the sample.

2. The comparative newness of petroleum geochemistry and consequently the amount of background data that has been built up over the years and is available to each type of analyst to enable him to interpret his results and provide the user with the required information.

The petroleum chemist can draw on data accumulated over the past 50-60 years whilst the petroleum geochemist has only data accumulated over the past 10-15 years.

Geochemical surveys are at their most useful when a broad survey over areas that are in the early stages of appraisal or exploration and when there is a need to upgrade existing information on applicable areas where decisions have been made regarding the possible extension or relinquishing of present sites. These may be the result of financial or political forces requiring rapid and above all, accurate decisions to be made.

In these circumstances there are at least two analytical areas where advances will have to be made for geochemical and associated methods of exploration. Each brings its own problems and its own solutions. One advancing area in marine exploration is that of using "sniffing techniques". These are dependant upon sampling the water above the seabed then analysing the water for traces of hydrocarbons and interpreting the results. It is necessary to have a more or less immediate interpretation of the results and this necessitates the use of immediate on site computers so that the survey equipment, including the survey vessel, may be precisely stationed and resample a suspect point without having tidal or other disturbances. It is essential to have good comparisons of the ratios of the various hydrocarbons obtained. It must be remembered that not all hydrocarbons found in the sea, especially in coastal waters, are results of decompositions leading from marine organisms to the production of oil and some of the light hydrocarbons such as methane and ethylene are the direct results of biogenisis from man related sources. It is essential to have accurate ratios of the hydrocarbons which are not man-made and relate these ratios to what may be expected to be present in an oil of industrial importance.

When such a survey has indicated that drilling is feasible, the geochemist's sample is usually a small amount of analyte heavily contaminated with a large amount of rock and the total level of analyte may be of the order of 1-2 mg. It is therefore necessary to consider what degree of accuracy can be expected, especially if the "client" requires, as is usual, the compositional information detailing the relevant amounts of the saturated hydrocarbons, the distribution of the n-alkanes and the isoprenoids; the aromatics and their relative amounts as well as the asphaltene content. The techniques which are presently used include molecular size distribution using gel permeation and characterisation of the various components by HPLC and by other chromatographic techniques often interfaced to various types of spectrometers including mass spectrometers. Since much of the earlier work on gel permeation and allied techniques for the allocation of molecular weights involved comparison techniques using, as standards, compounds which are not similar to naturally occurring hydrocarbons, generally with regard to shape, degree of complexity, there should be some fundamental research done using known hydrocarbons so that more accurate measurements can be made of compositions.

There will always be the problem of obtaining absolute measurements and it is conceivable that there will only be comparative methods available in the next ten years. It may require a quantum jump in the experimental philosophy to devise a technique which will give absolute values. In the meantime it may be appropriate to ask if absolute values are necessary or can we and should we carry on using the comparative values that we have to give us usable information.

ANALYSIS OF FRACTIONS FROM CRUDES.

From an analyst's point of view petroleum fractions must be regarded as some of the most complex mixtures known. They seem to consist of thousands of individual compounds, many of which have not been isolated from the various fractions. Physical separation of the various components is a very elaborate task and indeed for many of the heavier fractions it may not be possible with the present techniques.

It may be asked "Why, apart from a curiosity value, do we need to separate the various components?" and it is necessary to state that there seems to be no immediate reason for the knowledge that might be obtained if the identities of each of the individual constituent were known.

It is necessary to characterise the various fractions and to be able to indicate what compounds other than hydrocarbons are present in the fractions, since it is probable that these non-hydrocarbons have significant effect on the properties of some of the fractions. We are aware that the properties associated with some of these fractions, have sometimes been attributed to the presences of nitrogen and sulphur containing compounds.

One of the achievements of modern high resolutions mass spectrometry, with associated data handling, is the possibility of characterising many of the oil fractions by indicating the presence of sulphur, nitrogen and other elements having known isotopes abundances and masses which are capable of being differentiated. Developments in these areas will probably continue and will produce much potentially useful data.

An area of possible development is that which may arise when a combination of techniques such a new type liquid chromatographic separations of the fractions into smaller units followed by introduction of the separated units into the mass spectrometer using techniques such as discharger-assisted spray. This could be followed by chemical ionisation of the oils as a result of some of the solvents used in the separation and carried over into the mass spectrometer. This may seem fanciful, but there are always research workers, who are ready and capable of turning what may seem to be disadvantages. In this case, if it is not possible to remove the last traces of the chromatographic solvent, there may be a possibility of using this in an ionisation mode for obtaining better- that is- more easily used- fragmentation patterns.

There is little doubt that developments in the determination of the distribution of the various components of the n-alkanes and iso-prenoids by various types of chromatography will occupy many workers over the next few years.

The main problem will not lie in obtaining the results but in making relevant information from them. Sometimes we think that we have all the information and we try to make the results fit the information. This can lead to erroneous science. There is a good rule of thumb to follow in such cases; it is:- "Theory guide but experiment decides" and this is especially true when much of the theory is the result of perhaps to little experimentation.

PROBLEMS ASSOCIATED WITH ENHANCED OIL RECOVERY

It is not in the remit of the present lecture to discuss the reasons for injecting water into the wells to maintain reservoir pressure, but it is useful to indicate the possible size of one of the immediate problems, that of scale formation. Deposition of scale in formation channels is a wide spread source of problems in oil recovery.

It has been forecast that the amount of water which will be injected as filtered and treated seawater into the North Sea fields will be about 1.5 million barrels per day by the end of this year and will remain at this level for the next 5-10 years.

(Ref R.B.HOUGH. Paper presented at the symposium "Chemicals in the Oil Industry"; Manchester 1986 pages 1-12.)

There are two main types of scale: carbonate scale produced from the decomposition of the bicarbonates present in the formation water and sulphate scales which arise when a sulphate rich seawater is injected into a well whose formation water contains calcium, strontium and or barium. Barium sulphate scale is particularly tenacious in adhering to metal surfaces. The use of acid to remove carbonate scale, whilst not cheap, is relatively simple and efficient, but can give rise to corrosion problems associated with acid attack upon the metal surfaces. The sulphate scale gives much more serious problems and there are two methods of dealing with these:-

    (i)   The removal of exisiting scales,

and

(ii) The prevention of scale formation.

Whilst the removal of the formed scale by dissolution with various chelating and coordinating agents is commonly practised, the main area for research must be into methods of preventing scale formation.

There are at least three points of attack, each has its own proponents and whilst each has its own distinct problems, all have a common problem. This is simulating in laboratory conditions, the exact conditions which obtain the well at the time of the reactions.

Usually in research laboratories, certainly in most academic research laboratories, the work is done at temperatures between ambient and say 80°C and at ambient pressures.

Much of the knowledge of chemistry which we use has been obtained from work done under these conditions. There is little doubt that the thermodynamics of aqueous systems and mixed aqueous/ hydrocarbon systems, at temperatures well in excess of 100 degrees and at several atmospheres pressure, will differ from those obtaining in most general laboratories. Rates of reaction, diffusion and solubility are all important and are all altered.

Any work done in these areas of research for the petroleum industry must bear this in mind and whilst it may be justifiable to postulate from data obtained "on surface", the real test will be "In well" and it is probable that it will be necessary to have a large amount of philosophical thought devoted to a comparison of laboratory and field tests.

I consider that the three points of attack, all of which are at present under investigation, and which may give some reasonable chances of success are:

(i) Research into methods of affecting the morpholoy of the sulphate crystals so that they do not adhere to the surfaces. The phenomena of nucleation and crystal formation are some of the least understood in chemistry and the spin-off from such research could have very wide spread importance in many other industries.

(ii) The use of scale inhibitors, viz:- the use of substances which prevent the barium sulphate scale adhering to the surfaces;

and

(iii) The use of compounds capable of sequestering barium and hence preventing formation of barium sulphate.

There are two or three common criteria for any of the materials produced to prevent scale formation:-

(i) The material must be able to be used in very dilute form.

(ii) It must be able to be detected and determined, at exceedingly low concentrates.

(iii) It must be relatively inexpensive.

Scale inhibitors are usually carboxylate or phosphonate polymers, some of which are prepared <u>in situ</u> in the well by hydrolysis of suitable precursors. They are usually "administered" by squeeze treatment of the well. The timing of such treatment can be critical since if it is started too late the well would plug and if too early the chemical inhibitor would be wasted. Careful monitoring to detect initial injection water breakthrough at the production well is therefore essential in deciding upon squeeze treatments. Monitoring of the polyeletrolytes along the path is also essential and at the level of less than a few parts per million, obtaining an acceptable method of determination is proving to be a major problem for the research analyst. A breakthrough in this area is probable in the next few years, several workers, including some of the present author's team, are actively considering this problem. However, the media in which the work has to be done is not the most conducive towards success.

Although these polyelectrolytes are widely used, their action is not completely understood, even for flocculation or sedimentation of existing precipitates, and work in the present author's laboratories concerned with the study of these polyelectrolytes for use in precipitation and related phenomena, indicates that it is not yet possible to state which of the many theories, if any, is correct.

Another area of research is that of synthesising materials capable of sequesting barium and preventing the formation of barium sulphate. The electropositive character of the barium ion does not make barium a good acceptor for coordination.

EDTA and its analogues have been investigated but there seems little chance of a major breakthrough in this area. A more promising set of compounds are the CROWN ETHERS which can form complexes with the barium and also have sufficient hydrocarbon content to make the barium complexes reasonably soluble in the hydrocarbon part of the system. One of the advantages of these compounds is that they will dissolve already formed scale but a note of caution must be sounded, they are at present very expensive and the amounts required are relatively great. However, this area will probably be profitable in the future and it is a potential solution to the dissolution problem!

There is no doubt that the principle factors controlling scale formation in the oilfield are chemical in nature and a knowledge of the mechanisms of crystal growth and dissolution will go far in providing the answers to many of the scaling problems.

CORROSION

One of the areas being mentioned as being necessary for more research is that associated with corrosion and linked problems. Corrosion arises from a multitude of sources and among the major is the effect of oxygen as a corrosion enhancer. Scavengers are used on a very wide scale and the effects of these and any residual oxygen needs investigation. Whenever multithousand tonnes amounts of any expensive chemicals are used annually, then industry, with its eye on the "bottom line" requires more cost efficient materials to be produced.

There is a great deal of concern that the water injected to maintain the reservoir pressure will cause the production of large amounts of hydrogen sulphide as a result of the activity of the sulphate reducing bacteria, which contaminates the injected water.

The hydrogen sulphide, along with carbon dioxide enters the gas stream and produces an acidic gas which has deleterious effects on pipelines, burners

and apparatus associated with combustion. It is part of the work of the
petroleum-gas industry to ensure that the gas is rendered inert. Even
through a variety of processes for the "sweetening" of sour gas have been
used for many years, reports of new processes continue to abound. Many of
them are costly multistage processes and it is desirable to have the
minimum number of stages - i.e. a one stage process, whenever possible.
Research into processes which sweeten the gas stream and produce, as
by-products, useful sulphur containing chemicals, are due for
investigation. The sulphur is in a very reactive form and the chemical
energy should be used in a profitable manner.

THE PRODUCTION OF SECONDARY MATERIALS FROM OILS AND SHALES

The production of secondary materials from oils and related products of the
petroleum industry and the research that is needed in these areas forms so
large a topic, that it requires a symposium in itself. I, therefore,
intend to say nothing about this area except to commend it to any young
research worker in applied chemistry who wants to have a never ending
source of problems

PART 3.

ENVIRONMENTAL PROBLEMS

It must be accepted that monitoring only indicates when there is a given
level of material present, legislation decides whether or not there is
pollution.

There is no doubt that environmental pollution -of some type- will always
result from the petroleum industry. The type of pollution and the amount
will alter, it is hoped that the amount will decrease, but, nonetheless,
there will always be some pollution, which is the direct or indirect result
of the petroleum industry.

There are several ways in which the environmental aspects of petro-analysis
will provide problems in the future for the research petroanalyst; although
generally these will be developments of existing problems, there is always
a need to find a new, more economically acceptable, methods of monitoring
the effects of the petroleum industry on the general environment. It stems
from the need to have a better solution to the equation of the bottom line.
Although environmental monitoring is not a recognised way of increasing oil
production, it can produce information that will either allow more efficient
production of the starting product or it may help to provide information,
which will help in the task of bettering the public relations between oil
companies, their customer and the general public.

There is unfortunately a need for the oil producers to defend their
position against attack by the "environmental ecologists", there are oil
spillages, there will be oil spillages and there will be outcries that
these are causing a great deal of damage to the ecology. The methods for
"fingerprinting" oil traces are good and to a large extent they serve the
purposes for which they are intended. However, there is a need to ensure a
greater validity of the information that is deduced from the results.
Development of identification methods, probably using LC/MS or LC/FTIR
techniques with a greater amount of background data and information
retrieval, will be a definite source of research and development in the
next few years.

Future Trends in Petroanalysis

Atmospheric or gaseous pollution will also have to be considered and work on detecting and identifying sources of pollution must have a greater priority than it has at present. The areas involving FTIR will probably prove to be effective in helping to detect pollution at an earlier stage and so prevent further pollution.

The area concerned with pollution is an area where there is a need for education especially with regard to numbers. Concentrations of pollutants are usually measured in parts per million or even in parts per billion. The analyst generally quotes his results to an approximate figure, say 1-5 ppm or to the nearest 100 ppb. The fact that large expensive pieces of apparatus are used in these determinations seems to indicate to "clients", be they production manager, accountants or lawyers, that it should be possible to give more precise figures. I am sure that if one calculates what such a variation means, in terms that are more easily understood to the non-analytical public and can ensure that clients understand this in terms of their own familiar working units, then we may obtain a better understanding of the results and hence better cooperation.

A variation of 1000 ppb, in say two results from two different analysts, dealing with the concentration of a pollutant at the 1% level in the sea or in a fish liver, is the same as the variation of 1cm in 100m and it probably has the same significance to the ecological problem as a distance of 1cm has to average production manager or ecologist if they attempt to run 100 metres.

COSTS OF ANALYSIS AND SOLUTION OF THE EQUATION TO THE "BOTTOM LINE"

When we attempt to provide "as is analysis" it is generally found that accuracy and cost go hand-in-hand. The immediate from this is that money spent on analysis should not exceed what is necessary to ensure the information that is needed by the user. To ensure this is often possible to employ two sets of methods; the cheaper routine methods, which are used whenever possible and the more expensive methods which may be "reference" methods used in research oriented speculations.

It must be remembered that those who purchase analytical results usually get what they pay for, and as a general rule the clients are unwilling to pay very much.

G.F. Blundell (American Bureau of Standards 1933, (Ref:- Ind Eng Chem Anal END. Vol 5. 221, 1933) stated that buying analyses is similar to buying a suit. It is possible to obtain ready made mass produced suits which are relatively cheap since they are mass produced in relatively large quantities from a standard pattern and are quite satisfactory for the man whose figure follows conventional lines; the routine methods of analysis are similar to clothes, they are perfectly satisfactory so long as the composition of the material is not unusual and the ordinary accuracy suffices. The moment one departs from these specifications and enters the custom made department, the cost mount rapidly for both suits and for analyses.

A consultant or research analyst sells an intangible commodity -an opinion- and in many circumstances the client cannot see why he should pay a large amount of money for one opinion if someone elses opinion can be obtained for less than half the price. Care must be exercised if cost effectiveness is to be realised.

It must be accepted that often these opinions, unlike those given in litigation exercises, are not the result of exclusively mental processes;

the opinions are based on physical evidence as well as on a mental interpretation. This physical evidence needs to be obtained from and supported by, good physical resources. These physical resources will generally be regarded as costly. Whilst this may be correct, relative costs of equipment and of training must be compared. The most expensive piece of equipment in any laboratory is a poorly trained analyst.

In none of my speculations have I suggested that the machine will replace the man.

If the future trends in petroleum analysis are to result in good bottom line equations then industry will have to recognise that man training is an essential part of the equation. The cost of training is an essential part of the equation. The cost of training and the cost of good opinions are very small when compared to the cost of poor analysis and poor opinions.

The older professions and craft guilds had their own specialised training programmes, they did not rely simply on the established systems to ensure what they required they would get.

I am not suggesting that one of the future developments is to ensure that the petroleum industry works in the same way, but to ensure similar results it may be necessary to use similar methods.

The future of petroanalysis is well stocked with challenging problems for trained minds; BUT there are no future developments without trained minds, the petroleum industry cannot afford to ignore this and its implications.

Laboratory Integration: Automation and Robotics

G.B. Wilson
Esso Petroleum Co Ltd, Esso Research Centre, Nr Milton Hill, Abingdon, Oxon OX13 6AE

Summary

The Esso Research Centre, Abingdon, Chemistry Laboratories provide support to the Research and Development, Marketing, and Manufacturing Divisions of Exxon, both in the United Kingdom and abroad.

Over recent years a considerable amount of automation has been introduced. This paper highlights:-
* problems encountered
* issues addressed
* solutions employed
* some industry-wide implications

By recognising that many analytical tests contain similar processes (such as weighings, dilutions, and heating), it is possible to identify common solutions to problems encountered. Potential savings can thus be maximised.

This leads to the identification of Unit Operations (UOs) within any laboratory procedure. These UOs are frequently automated using task-specific apparatus, and if practicable a robot can be used to transport samples 'intelligently' between them.

Future developments lie in improving the 'intelligence' of the robot. Multiple-tasking, and knowledge-based systems are key features of the latest applications.

The benefits to both laboratory staff and the organisation include:-
* kudos
* improved quality
* reduced delays
* improved turnaround times
* increased capacity
* reduced 'in-lab' bottlenecks

At some time in the future, automation equipment, and robots in particular, will become as easy to use as the modern microcomputer and its' software. Robots will then become true 'end-user' automation tools. Many manufacturers would like to think that that time has already arrived. Yet, initial installations commonly take 1 year for a working chemist to implement. By adopting computing project management principles that effort can be substantially reduced.

Clearly higher pay-off tasks are automated first. Procedures which are less variable are the easiest to automate, often regardless of their complexity.

Automation calls for lateral thinking skills. Automating existing human activities can simply automate existing problems.

Changes in operation which may inconvenience customers, such as the enforced use of standard sample bottles, are readily accepted by them, when they recognise the possible benefits.

Automation changes methods. New standards must allow for alternative methods to be used, with the quality of the results being the determining factor. A quick route should be available for the validation of alternative methods to existing standards.

1. The Commercial Environment of the Laboratory

The Esso Research Centre, Abingdon, Chemistry Laboratories provide specialist product development and technical support for a broad range of petroleum-related product lines. Analyses are conducted on materials including; all types of lubricant, liquid transportation fuels, and LPG. Specialist support is provided to refining, transportation and marketing groups.

Most routine analyses are performed within the Analytical and Inspection Sections. Sample numbers depend on the particular analysis being performed, though typically Analytical Section will handle more than 1000 samples for each test procedure each year, while Inspection Section often conducts more than 10000 of each test.

With such large numbers of routine preparations and analyses it is not surprising that a considerable investment has been made in automation. The financial constraints affecting the decision to automate are constantly changing, and potential applications require regular reassessment.

2. Degrees of Automation

Manual Systems.

In many laboratories, automation exists as a number of discrete sets of apparatus performing specific test procedures. The apparatus is connected in such a way that a sample progresses through the analysis with the minimum of manual intervention. These complex arrangements of glassware, with associated plastic tubing conducting solvents do not usually involve electrical, electronic or microprocessor control.

Turnkey Systems.

The commonest form of instrumentation in the laboratory is the 'turnkey' system. Purchased as a standalone machine, these instruments perform a routine analysis or procedure, and usually produce reports which require further interpretation by the operator. Examples include X-Ray Spectroscopy, Atomic Absorption Spectroscopy, and Gas Chromatography.

Turnkey systems usually have self-contained data manipulation components, either based purely in electronics, or else provided by an on-board microprocessor. Where substantial calculations are involved most manufacturers of these systems will provide a dedicated computer to control the apparatus and produce the report of results. These computers are often

Laboratory Integration: Automation and Robotics

over-specified for the particular job in hand, and suitable software for the procedure will only be available on one type of computer.

Data Capture at Source.

When 'turnkey' machines are not available, the common approach to automation is to develop computing and electronic systems capable of capturing data from the apparatus involved, and storing it for subsequent analysis by computer. These solutions are generally developed in-house, though some manufacturers specialise in providing tools of this kind for specific procedures.

Where these systems provide for the transfer of data from the local microprocessor to a larger shared computer elsewhere, they are said to be 'Capturing Data at Source'.

Reprogrammable Devices.

In many cases samples are transferred, as well as data, between the stages of a complete analysis. When the degree of transport is too great for arrangements of tubing to suffice, a reprogrammable device is used.

Consisting of an 'arm', with various 'hands' capable of different functions, these act under the control of a computer and an arrangement of electronic interfaces. Typically, a system of this kind will collect a sample from a tube in a rack, extract sub-samples, add solvents in measured quantities and suitably mix the components. The device may then feed an automated instrument, collect and weigh fractions as they appear, and provide some manipulation of results before producing a report.

Reprogrammable devices are used so that changes in procedures can be made at a much lower cost than for custom-built, data capture or turnkey systems. Similarly, if a turnkey system becomes available later, the existing apparatus can be re-used elsewhere, if it is economically viable to invest in the turnkey unit.

Another name for reprogrammable devices is Robot, a term derived from the Czech 'robotnik' meaning 'serf'. These systems have a popular 'anthropomorphic' image. In reality, laboratory robots usually consist of either a static base with a revolving arm, or a track- or gantry- based arm, again revolving about the base. To avoid misconceptions, the more cumbersome, but accurate phrase 'reprogrammable device' is often used.

Remote Job Scheduling.

With a reprogrammable device performing an analysis, and the data generated being captured at source, it is a natural progression to transfer information in the opposite direction.

A daily list of chores can be transferred from a host computer, providing a Laboratory Data Management System, to the microcomputer controlling the robot. Before embarking on any activities, microcomputer-based software can check an inventory of replenishables, and can alert laboratory staff if certain materials are needed.

Under the control of a flexible-response computer program, the robot begins the first task of its workload, often the longest to perform. The equipment continues this task until it reaches a long pause; for example, waiting for a digestion phase to complete. The software identifies the next most critical task, and the robot begins work, returning to the original job, when safe and appropriate to do so. This process is a form of multitasking.

Activities for multitasking need not be confined to the robots workload, they can include items of routine maintenance, such as lubricating a track.

3. The Integration Concept

With so many different working systems, there is an inevitable need for communication between them. It may be necessary for specimens to be exchanged; for prepared samples to pass between instruments; for data about specimens, samples or tests to be transferred; for job control information to be transmitted; and not the least for the chemists operating instruments to move between machines.

This is why a futuristic laboratory is conceivable in which everything from specimen receipt and the request for testing, to the reporting of validated results is connected together.

Integration is the connection of the discrete activities in a laboratory, so that they operate synergistically.

The emphasis throughout Laboratory Integration lies in ensuring ease of connection between all elements of Automation. In particular, in maintaining common interfaces to all components, whether between computers and instruments, operators and instruments, or instruments and apparatus.

4. Prioritising Integration

Commercial Objectives

Throughout any integration process there are two important constraints; the quality of the end results, and the commercial viability of automation.

Having identified possible applications for automation, it may not be economic to proceed with automation at this time. A clear Integration objective ensures that all activities will ultimately contribute to the whole, rather than frittering valuable resources.

Priorities

Three principal criteria for prioritising automation have been identified. When a particular application receives low priority, it is important that it is monitored, as changes in the availability of alternative technology can radically affect its importance.

To establish priorities, the Analytical Method is considered in 7 stages, as described later. Encouraging consistent means of automation, these reflect the differences in technology used to handling specimens, samples, data and results.

Standardisation

Activities which are to be automated need to be limited in their variability.

Often Specimen Receipt and Handling are the hardest to standardise, though the introduction of bottling and labelling conventions can be extremely effective precursors to automation.

Conversely, the presentation of results, involving only software development, can be the easiest stage to automate.

Technical Skills

Ideally, all staff would perform work which was varied, stimulating, and demanding. Automation can make two contributions in this area. Firstly, it can free individuals from the need to carry out dull, repetitive tasks. Secondly, it can introduce them to more demanding skills involving new technology.

Until now, existing technology has favoured the automation of technical tasks, often involving sample preparation and handling, and data collection.

Developments in mechanical aspects of automation will favour the provision of systems receiving and handling specimens. Meanwhile improvements in information technology, and especially concerned with expert systems, will allow further automation of routine decision making processes.

Monotony

Certain tasks involve a considerable amount of work which is both repetitive, and apparently without reward, yet is crucial to the satisfactory completion of a job. Where these activities are concerned less with manual dexterity, than with controlled physical manipulation, they can often be automated using a robot.

Seen in this light some activities within the laboratory, such as glass washing take on a new significance, in that they are often regarded as the least rewarding, yet they are the hardest to automate.

Monotonous activities occur throughout the analytical method, and individual cases have to considered independently.

5. Applying Automation

The Analytical Method

Most laboratory procedures, at least at the Esso Research Centre, can be resolved into 7 stages.

- i)   Specimen Receipt and Distribution
- ii) Specimen Handling
- iii) Sample Preparation
- iv) Sample Analysis
- v)   Data Collection
- vi) Data Processing
- vii) Result Reporting

The origination of requests for work falls outside this scheme, though it is important in the concept of Integration. It cannot always be assumed that such requests occur before the specimen is received.

Frequently, small automation projects associated with a particular laboratory will be applicable in other laboratories. This is because of similarities in the procedures followed at any one stage in the analytical method. Thus a computer module designed for the processing of calibration data may be applicable to many, otherwise discrete, systems. Through adopting Integration standards, a single solution can be applied throughout, without the need to 'reinvent wheels'.

Operational Units

In the systematic analysis of existing methods, and the design of computerised alternatives, there is a need to recognise 'operational units'. These are the specific activities which are performed by an individual during a procedure.

By analysing the flow of information from one activity to the next it is possible to reduce a computerised solution into the minimum number of steps necessary to achieve the same end result.

For automation systems there is an additional need to reduce the physical transfer of materials (specimens, samples, and replenishables) to a minimum.

By studying a number of fully operational systems, it is possible to estimate the time taken by a typical unit operation, and hence to estimate the time it will take for the complete automated procedure. Laboratory Unit Operations include such activities as transferring a tube from one position to another; weighing a test-tube; or dispensing 50ml of solvent.

It is only by completing this type of study before any attempt is made to automate, that the feasibility of particular solutions can be assessed.

Though such estimates are prone to substantial error, they serve to :-
* highlight probable bottlenecks in operation;
* allow quantitative choice between alternative systems;
* reduce subsequent development time.

6. Implementing Automation

Team Structure

Even the smallest automation projects benefit from a team approach to implementation. Four key roles have been found to be needed.

The involvement of the people who are going to use the final system is vital. Firstly, in gaining their commitment, and secondly in tapping their knowledge of the working methods. Systems which are developed by one individual working alone, seldom meet the day-to-day needs for which they are intended.

For this reason the method of Project Management adopted by Esso, calls for the approval by these users, at several crucial stages before completion.

A systems analyst/programmer working to industry standards (generally, SSADM, the Structured Systems Analysis and Design Methodology), ensures that an application which has been developed by one group can be easily supported and updated by others. In this way, subsequent integration with other systems, at all levels, requires minimal effort.

The automation engineer, also working to exact standards, is responsible for the design and development of the hardware associated with a project. The advantage of an individual who is not responsible for the day-to-day operation of the procedure, lies in the ability to see alternative working methods. As described below, this is essential if the automation of existing problems, sources of error, and inefficient practices, is to be avoided.

Finally, a Project Manager responsible for the initial consideration of possible projects, and the administrative activities associated with a number of ongoing applications, can overcome excessive interruptions to the software and hardware specialists. One of the most significant delays in implementation comes from the distraction of these individuals to other problems!

Project Management

For some years Esso has employed a 6 stage method of Project Management for the implementation of computing solutions. This ensures that a continuity can be achieved during development of large scale projects, and the ease of integration and maintenance of all projects.

'Scoping' and 'Exploration' set out the exact nature of the problem to be addressed. 'Specification' and 'Design' are performed by the software and hardware engineers, and spell out what must be done, what it will look like, and how much it will probably cost. Along with the final two stages, 'Implementation' and 'Handover', this provides six points at which the end-users are involved in the testing and validation of the new method.

By adopting this approach considerable savings in development time can be achieved, when compared with projects developed by dedicated users working alone. Furthermore, the end product requires less maintenance and can be incorporated into larger systems should the need arise.

Strategic Planning

The need to communicate between systems at a number of distinct levels in their organisation has been recognised for some years.

The evolution of standards, such as the IBM P.C. among microcomputers, was initially left to market forces. With increasing competition from other manufacturers, no single producer can afford to develop machines or software which are not compatible with the emergent standard.

With experience, larger organisations are insisting not only on hardware and software compatibility, but also on common standards for the development of these products.

For development, SSADM (Structured Systems Analysis and Design Methodology), has been adopted by many, especially within the Defence industry. For connectivity, a seven layer model, known as OSI, has been adopted by Government and commercial organisations.

OSI, or Open Systems Interconnection, attempts to define standards for the transmission of electronic information, data which controls systems, addressing of data between systems, dialogue, screen formats, and the connection between electronic and mechanical components.

The first standard to directly affect automation was MAP, or Manufacturing Automation Protocol, developed mainly by General Motors.

There are no standards, defined or implicit, for the connection of Laboratory Automation systems. Yet most Laboratory Automation systems consist of a number of discrete automated components, gathered together for that particular application. With increasing automation there is a need to share apparatus, data and operators. Without standards careful scrutiny prior to purchasing, will become increasingly important.

For smaller suppliers this will mean that the larger corporates' buyers will become increasingly reluctant to invest in speculative purchases. Until recently, larger manufacturers had appeared exempt. However, they too report substantially reduced profits, as products are bought for their performance and technical qualities, rather than relying on the commercial integrity of the larger organisation.

7. Implications of Automation and Integration

Changing Operating Conditions

The automation of existing manual methods is often inefficient. There may be particular tasks which are harder to perform than others, and in which the degree of accuracy is likely to be reduced. There may be tasks which are only carried out because of the laboratory layout.

By automating a procedure step-by-step, the difficulties and problems which occur as a consequence of passing from one step to the next, are merely automated too.

In practical terms, this highlights the need for individuals, with a particularly lateral approach, to analyse and design systems. It also implies that operating conditions and test procedures will change as a natural consequence of automation.

Introducing New Procedures

Unfortunately, many test procedures which have been in use for a considerable period of time, are based on specific apparatus, and require, clearly defined, manual activities.

Automation may render many of these redundant, and will require speedy validation of new methods. Automation within different laboratories is also likely to differ, as computing systems may not be the same, and as considerations of space vary.

There will therefore be an increasing need for the validation of results without regard to the test procedure used. Furthermore this validation will need to be quick, as companies investing in automation will be reluctant to delay its implementation, whilst waiting for committee approvals.

Validating Results

Provided that the automated method does NOT use alternative theoretical bases for analysis, the procedure adopted should not affect the accuracy of the reported result.

When adopted procedures ARE based on alternative theory, and once that theory is accepted for practical applications, the only method of comparison between systems is that of the precision of the final reported results.

Consequently, with the introduction of further automation there is a need for test procedures to cease to be defined by method, and instead to be based on quality and precision parameters.

8. Acknowledgements

The application of robotics at the Esso Research Centre, Abingdon, has been the result of considerable persistence, and highly innovative work by two individuals, Dr. Ted Wright and Mr. Roland Bustany. I should like to thank them for their help in all aspects of this project.

Discharge—Thermospray Ionisation of Lubricant Components

M. Thompson
Shell Research Centre, Thornton Research Centre, P.O. Box 1, Chester CH1 3SH

1. INTRODUCTION

The combination of high performance liquid chromatography with mass spectrometry (LC/MS) is a very powerful tool for the analysis of labile and involatile components in complex mixtures. However, in contrast to the more mature combination of GC/MS, a wide range of interfacing techniques are used. Commercial interfaces range from transport (moving belt) systems, where virtually all of the solvent is removed, to direct liquid introduction (DLI) without any pre-concentration. Each of these has limitations and no single interfacing technique is applicable to all sample types and solvent systems. The thermospray (TS) interface[1] is particularly attractive as a source of ions for MS, being simple in design and able to accept high solvent flow rates (up to $2 cm^3 . min^{-1}$). The latter feature means that, unlike DLI, direct coupling to HPLC is possible and sensitivity need not be sacrificed by flow-splitting.

TS operates by producing a supersonic jet of liquid droplets from a solution as it emerges from a heated capillary tube. This solution contains the analyte along with a volatile buffer such as ammonium acetate. As the liquid is nebulised, random sampling of the ionic species in solution causes some of the droplets to become electrically charged.[2] Evaporation then produces cluster ions which undergo ion-molecule reactions if favourable pathways are available.[3] This mechanism can only operate under certain conditions and consequently there are several problems associated with conventional TS which limit its range of applications.

(i) The buffer is essential to the TS process and, although it can be added post-column if it interferes with the separation, this restricts the choice of mobile phase to reversed-phase solvents.

(ii) Sensitivity varies greatly with compound type, and non-functionalised molecules, such as polycyclic aromatic hydrocarbons, do not ionise at all.

(iii) Generally, pseudo-molecular ions are produced with little or no fragmentation and consequently a lack of structural information.

(iv) The mass spectral peaks tend to be noisy which makes accurate mass measurement, and hence determination of atomic composition, rather difficult.

Particularly as a consequence of (i) and (ii), most applications of TS have been in the analysis of polar biological molecules. In the petroleum industry, where a much wider range of structures and polarities are encountered, transport interfaces have been the more frequent choice.

Some of the problems of TS can be overcome by using an electron gun to enhance the ionisation, but filament life tends to be short because of the high solvent pressure in the source. An alternative method of ionisation is to create an electrical discharge in the solvent vapour,[4] which gives a similar increase in sensitivity to that obtained using an electron gun. This approach overcomes the need for a buffer and consequently a greater range of solvents can be used. An additional advantage is that the solvent jet velocity can be reduced by using a wider inlet capillary, which is less likely to block than the < 0.1 mm bore tubing required for conventional TS.

Commercial discharge-thermospray (DTS) ion sources differ slightly in their construction, but most use a pointed discharge electrode somewhere downstream of the capillary inlet (Figure 1). An alternative design uses the metal inlet capillary as the discharge cathode; this ensures that all of the

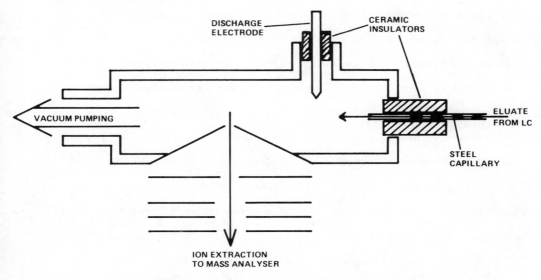

FIG. 1 — LOCATION OF DISCHARGE ELECTRODES IN THERMOSPRAY ION SOURCE

solvent plume passes through the cathode dark space, where the greatest potential gradient exists, prior to entering the glow discharge region.

2. EXPERIMENTAL

The following work was carried out in order to assess the applicability of DTS to the type of compounds encountered in the analysis of lubricants. The influence of various experimental conditions on the type of spectra obtained, for samples of different structure and polarity, was investigated. These variables included source geometry, discharge current, sample concentration and solvent. The samples included saturated and aromatic hydrocarbons/hetero species, representing petroleum base-stock components, and various functional additives such as antioxidants, antiwear/extreme pressure additives and metal deactivators.

Experiments were carried out on DTS sources with both discharge geometries, i.e. capillary electrode and point-discharge types. These were mounted on double-focussing magnetic mass spectrometers.

HPLC separation conditions could not be investigated until the compatability of various solvents with DTS was known, and hence on-line separation was not attempted at this stage. Instead, 10µl aliquots of approximately 1% w/v solutions were introduced, via a loop injector, into the solvent stream flowing at $1.5 \text{cm}^3 \cdot \text{min}^{-1}$. Choice of solvent was dictated by the solubility of materials to be analysed and previous experience with stand-alone HPLC on some of these. The solvents investigated were acetonitrile, acetone, dichloromethane (DCM) and dichloroethane (DCE). All of these gave DTS spectra for most of the compounds studied, but the first two tended to give a wide range of background ions and complex clusters with the analyte ions. The chlorinated solvents gave very good results and most of the work concentrated on these. It was discovered during the experiments that the DCM was contaminated with a small amount of methanol (< 1%). This resulted in some important differences from DCE which will be discussed later.

3. RESULTS AND DISCUSSION

3.1 Effect of discharge current

Studies carried out by the interface manufacturers, on polar biological molecules in reversed phase solvents, have shown that varying the discharge current can alter the degree of fragmentation. Low currents tend to yield almost exclusively molecular ions whereas high currents give fragmentation similar to that observed in electron impact (EI) spectra.

For the solvents used in these experiments a stable discharge was only obtained over a limited current range, corresponding to a potential difference of about 600-800V. No significant changes in the spectra were observed in this range.

3.2 Effect of source geometry

The only case where mass spectral differences attributable to discharge geometry were observed between the two sources was for a saturated hydrocarbon (n-tetracosane). With the capillary electrode interface, almost exclusively [M − H]$^+$ ions were detected (Figure 2), while with the point discharge system EI type fragmentation was observed (Figure 3). The spectrum obtained using the latter interface differed from EI in that [M − H]$^+$ rather than the molecular ion was the highest mass, and a different distribution of fragment ions was observed. This example shows that passing through the high potential gradient of the cathode dark space, in the case of the capillary electrode system, does not give increased fragmentation as might be expected. One reason may be that most of the analyte molecules are protected by solvation in the discharge region of this interface.

3.3   Mechanism of ionisation

The high solvent pressures and long residence times in the TS ion source result in various ionisation mechanisms. The relative intensities of the various types of ions detected is strongly dependent on the structure of the sample molecules.

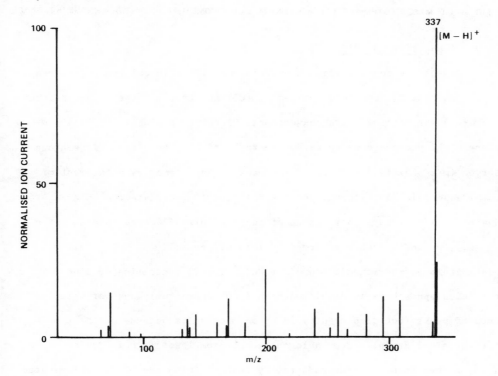

FIG. 2 — DTS MASS SPECTRUM OF TETRACOSANE IN DICHLOROETHANE (CAPILLARY ELECTRODE INTERFACE)

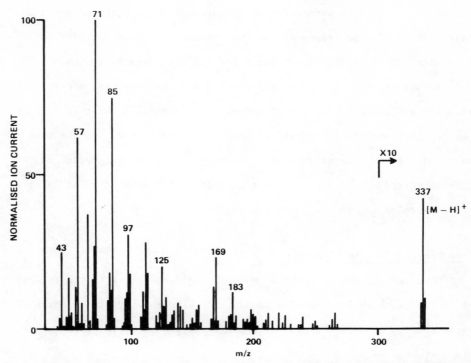

FIG. 3 – DTS MASS SPECTRUM OF TETRACOSANE IN DICHLOROMETHANE (POINT DISCHARGE INTERFACE)

3.3.1 Negative ion DTS

Negative ion DTS spectra of organometallics in chlorinated solvents closely resemble Cl^- chemical ionisation (CI) spectra.[5] The upper spectrum in Figure 4 was obtained with conventional Cl^- CI of a zinc dipropyldithiophosphate using carbon tetrachloride. The highest mass group observed (m/z 525) is due to pseudo-molecular ions formed by attachment of Cl^-. The anionic dialkyldithiophosphate ligand (L^-) is detected as a fragment at m/z 213. A third group of peaks at m/z 347 correspond to $[M + Cl_2 - L]^-$, which is probably formed by loss of neutral dialkyldithiophosphoric acid from $(M + HCl_2^-)$. DTS ionisation of zinc dibutyldithiophosphate in dichloroethane (lower spectrum) gave very similar groups of peaks (but displaced by 28 or 56 mass units because of the different alkyl group). The analogous copper complexes form molecular anions, rather than chloride adducts, by both Cl^- CIMS and DTS in chlorinated solvents.

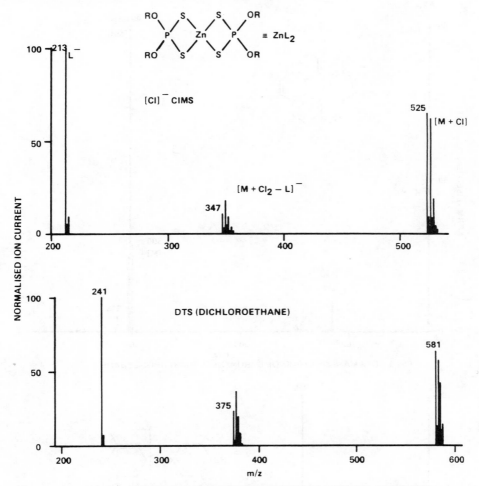

FIG. 4 — MASS SPECTRA OF ZINC DIALKYLDITHIOPHOSPHATES BY CHEMICAL IONISATION AND DISCHARGE–THERMOSPRAY

3.3.2 Positive ion DTS

Positive ion DTS spectra of aromatic molecules contain peaks due to both molecular and protonated molecular ions. Very little fragmentation is observed from stable structures such as coronene (Figure 5). Fragmentation of more labile structures is generally different from that seen in 70eV EI spectra, but the peaks can be easily assigned and are valuable for structure elucidation, as for example for the hindered bis-phenol AN-2. (Figure 6). Formation of $(M + C_2H_4Cl^+)$ ions was observed for most samples in DCE, e.g. dibenzothiophene (Figure 7), but the presence of a trace of methanol in DCM

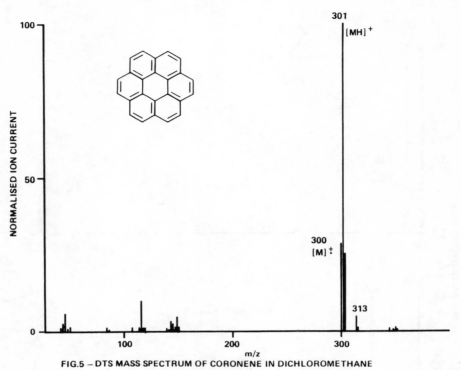
FIG.5 – DTS MASS SPECTRUM OF CORONENE IN DICHLOROMETHANE

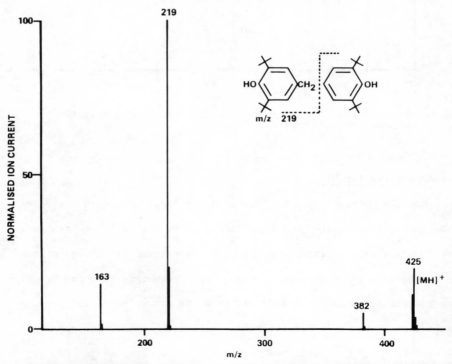
FIG. 6 – DTS MASS SPECTRUM OF 3,3', 5,5'–TETRA– t–BUTYL–4,4'–DIHYDROXY–DIPHENYLMETHANE (AN–2) IN DICHLOROMETHANE

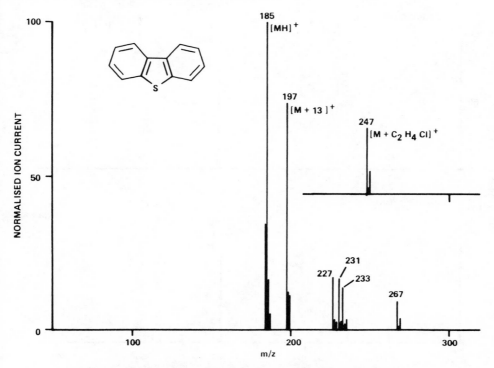

FIG. 7 — DTS MASS SPECTRUM OF DIBENZOTHIOPHENE IN DICHLOROMETHANE
(INSET = DICHLOROETHANE)

gave a more complex series of adduct ions. Peaks at M + 45 and M + 79 were observed in most cases and assigned to addition of chloromethyl ether ions, $CH_3\overset{+}{O}=CH_2$ and $ClCH_2\overset{+}{O}=CH_2$ respectively. Varying amounts of $(M + 13)^+$ ions were also detected, the corresponding peak being particularly intense in the spectrum of dibenzothiophene. It seems probable that these ions result from fragmentation of $(M + CH_2Cl^+)$ ions.

3.3.3 Effect of sample pressure

At relatively high sample pressures, the probability of reactions between sample ions and neutral sample molecules may be sufficiently great that these compete with sample/solvent ion reactions. The positive ion spectrum of didodecyl disulphide in DCM was found to change as the sample pressure in the source varied across a loop injection peak (Figure 8). There

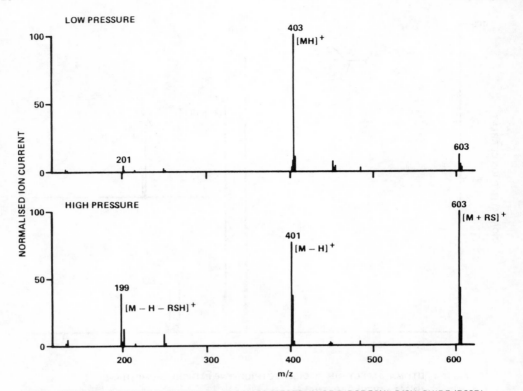

FIG. 8 - EFFECT OF SAMPLE PRESSURE ON DTS MASS SPECTRUM OF DIDODECYL DISULPHIDE (RSSR)

are several possible mechanisms which could account for these changes. At low sample pressures, solvent ions are the predominant CI reagent, and protonated molecular ions (m/z 403) give rise to the base peak. As the sample pressure increases, these ions disappear as RS^+ fragment ions react with the sample molecules to produce ions at m/z 603, which then fragment by consecutive loss of RSH to m/z 401 and 199. This is illustrated by the ion current profiles (Figure 9) in which m/z 403 (MH^+) maximises on the leading and trailing edges of the total ion current (TIC) trace, but m/z 603 ($M + RS^+$) and the fragment peaks at m/z 401 and 199 show maxima in the valley of MH^+. Alternatively, $2MH^+$ ions may be produced as short lived (and consequently undetected) intermediates which fragment by loss of RSH to give m/z 603, 401 and 199.

Similar effects were observed for the metal complexes, where high sample pressure resulted in formation of $[ZnL_3]^-$, $[Zn_2L_3]^+$, $[Cu_2L_3]^-$ and

Discharge - Thermospray Ionisation of Lubricant Components

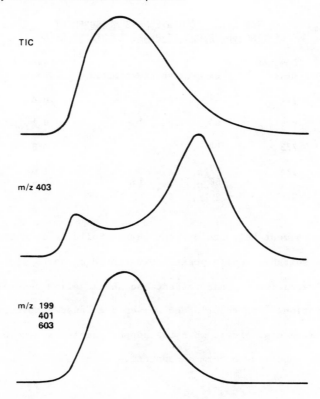

FIG. 9 – POSITIVE ION CURRENT PROFILES FOR LOOP INJECTION OF DIDODECYL DISULPHIDE IN DICHLOROMETHANE

$[Cu_2L_2]^+$. Change of oxidation state in copper complexes, as shown in the latter two ions, is also observed in CI and fast atom bombardment spectra.

3.4 Mass measurement accuracy

The exact masses of ions in the protonated molecular ion cluster for zinc dioctyl dithiophosphate were measured, at a resolution of about 5000, by scanning the accelerating voltage and peak matching to polyethylene glycol ions. The mass measurement accuracy was better than 2ppm (Table 1). Magnet scanning mass measurement accuracy is usually about 5-10ppm.

3.5 Stability and lifetime of the DTS sources

The lifetime of this type of ion source is generally limited by fouling of the ceramic insulator which isolates the discharge electrode. Use

TABLE 1 - ACCURATE MASS MEASUREMENT OF $[ZnC_{32}H_{69}P_2O_4S_4]^+$ PEAKS IN DTS SPECTRUM

Nominal Mass	Calculated Composition	Error (ppm)
771	^{64}Zn, $^{12}C_{32}$	+0.4
772	^{64}Zn, $^{12}C_{31}$, ^{13}C	-0.1
773	^{66}Zn, $^{12}C_{32}$	-0.8
774	^{67}Zn, $^{12}C_{32}$	-1.0
	^{66}Zn, $^{12}C_{31}$, ^{13}C	+1.9
775	^{68}Zn, $^{12}C_{32}$	-0.8

of water as the solvent helps to prevent carbon build-up from the samples, but is rarely a suitable mobile-phase. Oxygenated organics such as methanol are the best alternative in this respect and should allow several days continuous operation. The chlorinated solvents allowed continuous operation for more than six hours, except when the copper complexes were being analysed. These appeared to form a conductive deposit which shorted out the insulator.

3. CONCLUSIONS

1. DTS ionisation is applicable to a wide range of polar and non-polar compounds, including many which are not amenable to conventional TS.
2. Chlorinated solvents give good CI-type spectra in both positive and negative ion modes.
3. The observed fragmentation is relatively simple and can be easily correlated with structure.
4. Mass spectral peaks are well defined and can be mass-measured accurately.
5. Source operation is stable so long as the insulator can be kept clean.

4. REFERENCES

1. Thermospray interface for liquid chromatography/mass spectrometry : C.R. Blakley and M.L. Vestal; Anal. Chem. $\underline{55}$, 750-754 (1983).

2. Statistics of liquid spray and dust electrification by the Hopper and Laby method : E.E. Dodd; J. Appl. Phys. $\underline{24}$(1), 73, (1953).

3. Studies of ionisation mechanisms involved in thermospray LC-MS : M.L. Vestal; Int. J. Mass Spectrom. Ion Phys. $\underline{46}$, 193-6 (1983).

4. Discharge ionisation as a secondary ionisation method for use with thermospray : C.H. Vestal, D.A. Garteiz, R. Smit and C.R. Blakley; Proc. 33rd Annual Conference on Mass Spectrometry and Allied Topics. San Diego. (1985). p771-772.

5. Selective chemical ionisation using chloride ions : R.P Morgan, C.A. Gilchrist, K.R. Jennings and I.K. Gregor; Int. J. Mass Spectrom.Ion Phys. $\underline{46}$, 309-312 (1983).

Advances in Mass and N.M.R Spectroscopies to Characterize Heavy Petroleum Fractions

M. Bouquet, D. Jolly, A. Bailleul and J. Brument
Total-France, B.P. 27, 76700 Harfleur, France

Sophisticated analytical techniques such as nuclear magnetic resonance (NMR) and mass spectrometry (MS) may be used today as high level routine investigation tools, allowing to obtain rapid, quantitative and exhaustive information dealing with heavy petroleum cuts.

This target was reached successively in NMR and MS by several setting up and checking steps in each technique, starting with, respectively, well-known NMR sequences and basic computerized MS calculation matrixes and adding home-made modifications based on standards or reference samples obtained from round robin tests.

Today these techniques are routinely applied to feed and product characterization, allowing for a better understanding of refinery processes mechanism (follow up of hydrogen enrichment through hydrotreatment processes, prediction of coke formation in FCC, follow up of changes in crude oil constitution, etc.) and providing a providential alternative to sample consuming determinations like cetane number or molecular weight.

NMR

Successive developments on NMR techniques have been performed recently, and several improvements have been successfully applied to petroleum analysis field.

Based on well-known sequences, a home-made ^{13}C NMR routine method has been built up for characterizing and measuring structural hydrocarbon units in any petroleum cuts, including heavy fractions (i.e. from gasoline to asphaltenes), and this without any assumptions on chemical shifts to elucidate the spectra.

Inverse gated decoupling and spin echo experiments with relaxation reagent, followed by an appropriate computation of data allow us to generate an accurate and quantitative estimation of CH_n groups proportions. This work has been published recently elsewhere (1) (2).

The spectra edition provides structural data which are translated through appropriate computation into so-called "structural hydrocarbon patterns". Vacuum distillates, bitumens and deasphalted oils are routinely examined by this technique ; we derive from these analysis characteristics like : aromatic ring index, C/H ratios, $C_A/C_P/C_N$ balance, etc.

These new analysis find applications in several fields :

- yield structure and product properties prediction for conversion (FCC) on hydrotreatment processes,

- products characterization like cetane number, which is routinely done by ^1H NMR analysis for all of our gas oils.

In the cases of FCC, for example, ^{13}C NMR technique is used to identify the precise breakdown of the feed atoms into aromatic, naphtenic and paraffinic species : $C_A/C_P/C_N$ for FCC feedsotcks. C_A is then correlated to gasoline production through the process. Carbon to hydrogen atomic ratios for the aromatic and aliphatic structures are easily estimated and provide pertinent indexes for cracking prediction of heavy feedstocks. Even, the coke production formerly predicted from the CC of the feed can much better foreseen by using an NMR parameter derived from the "hydrocarbon structural pattern" of aromatic structures.

This work has been published recently elsewhere (3)(4).

Molecular weight of the aromatic fraction of very heavy cuts (i.e. vacuum residue) may be derived : this determination has no experimental limit and is in good agreement with similar determinations performed with other new techniques.

The "structural hydrocarbon patterns" get their full significance when they are expressed in carbon numbers and allow a realistic comparison between heavy residues.

MS

More than ten years ago, Hood, Fisher and Gallegos separately developed MS techniques and matrix calculations in order to determine the chemical constitution of heavy cuts, preliminarily separated into aromatics and saturates.

The main improvement we introduced was to perform a Fisher type MS analysis on the whole cut, avoiding a time consuming separation by liquid chromatography (1).

Our method may be applied to a wide variety of hydrocarbons cuts, ranging from gas oils to vacuum distillates with final boiling point up to 525°C. Reliability has been established by comparison with other techniques applied to round robin test samples. The typical data sheet describes an Arzew 230-340°C gas oil, standard cut obtained by cooperative studies between French refiners.

Another improvement has been set up by using the data obtained from our high voltage Fisher type analysis to establish a quantitative carbon numbers distribution of each aromatic family of the analysed petroleum cut. Meanwhile an average molecular weight of each aromatic family can be estimated as well as an average molecular weight of the whole analysed cut.

This technique has found application in several field of our research in petroleum refining, as demonstrated by the following examples :

- Follow up of constitution drift for crude oils.

We provide a set of results dealing with an heavy vacuum distillate from Alwyn crude.

All the aromatic families are detailed by carbon number distribution and a final average molecular weight is estimated. This kind of result has been checked by NMR technique.

- Follow up of hydrogen enrichment or chemical changes of FCC feedstocks or products through hydrotreatment, as we may observe on the LCO hydrotreatment and hydrodesulfuration results table we give herein.

- Accurate cetane number estimation for small samples of either SR gas oil or LCO, saving time and sample consuming engine determination. In this later case, the same determination can be achieved by ^{1}H NMR as well (4)

Conclusion

With specific adjustments and appropriate calculations NMR and MS have become power full tools for precise structural analysis of petroleum cuts.

When applied to feeds and products characterization of conversion or hydrotreatment processes, these techniques allowed for significant improvements in processes comprehension as well as yield structure and product quality prediction.

But a new step is reached when these techniques are run to give structural information and data to be used in prediction models in order to foresee products quality, and avoid expensive and time consuming petroleum tests and pilot assays.

We may think that a future step may be reached when we'll succeed in obtaining good prediction of quality products based on raw determinations (density, viscosity, etc.) we try to correlate with NMR and MS sophisticated parameters.

References

(1) International Symposium on Characterization of Heavy Crude Oils and Petroleum Residues, Lyon, June 1984, Technip Editions

   M. Bouquet, J.C. Roussel, B. Neff : page 322 ; D. Joly : page 416.

(2) Fuel, 1986, Vol. 65, September, page 1240

   M. Bouquet and A. Bailleul.

(3) Oil and Gas Journal, 1986, September 15, page 95

   M. Denmar, A. Triki, J.P. Franck.

(4) World Petroleum Congress, March 1987, Houston, topic 18, PD, Paper 1

   J.L. Mauleon, J.B. Sigaud, J.M. Biedermann, G. Heinrich

The Analysis of Volatile Organic Compounds using Thermal Desorption/GLC/MS

P.J.C. Tibbetts, A.J. Holland and R. Large
M-Scan Limited, Silwood Park, Sunninghill, Ascot, Berks SL5 7PZ

INTRODUCTION

The analysis of complex mixtures of organic vapours (gases) has long presented the petroleum analyst with a major challenge. Early MS gas analysis proved highly specific for simple mixtures, but the method lacked both sensitivity and dynamic range and for complex mixtures was largely superseded by GLC, a method with high separation efficiency. The GLC method, however, lacked specificity; the assignment of peaks often required laborious calibration of retention data with authentic materials.

Neither MS or GLC was suitable for the direct analysis of trace atmospheric contaminants. In such cases it was necessary to adopt some form of sample enrichment, such as concentration onto activated charcoal, the adsorbed volatile organic compounds being removed by solvent elution and analysed by GLC or GLC/MS. This method has been used extensively in the occupational hygiene field, but is unsuitable for low boiling compounds, which may not be resolved by GLC from the extracting solvent.

The majority of the above limitations are removed by concentrating organic vapours onto an adsorbent such as Tenax TA and analysing the resulting concentrate by thermal desorption (TD)/GLC/MS. The method is applicable both to the analysis of atmospheric or headspace samples, where air is drawn through a pre-conditioned Tenax tube, and to the analysis of aqueous samples, where volatile compounds are removed from water by purging with helium and collected on Tenax. Observed components are assigned by automatic library matching.

The TD/GLC/MS method has been developed by ourselves into a routine analytical tool. Examples of its power and flexibility are given below.

DISCUSSION

The limitation of mass spectrometry (MS) for the analysis of organic vapours (gases) is shown clearly in Figure 1. The EI mass spectrum of gas separated from a crude oil is dominated by intense hydrocarbon fragment ions (m/z 27, 29, 41, 43), which are common to the various aliphatic components present. There is little obvious evidence for molecular ions. The degree of specificity is therefore low.

In certain cases individual components with distinctive mass spectra can be detected by direct MS. A cyclic C_6 alkane (molecular ion: m/z 84) is apparent in the mass spectrum of contaminated air (Figure 2). Combined gas liquid chromatography/mass spectrometry (GLC/MS) analysis of the same sample, however, establishes that the cyclic C_6 alkane is in fact a mixture of cyclohexane and methylcyclopentane and reveals the presence of at least 11 additional components. The majority of these have been identified by MS library matching (Figure 3). In this respect there is an obvious advantage in separating the individual components of the mixture by GLC before MS analysis.

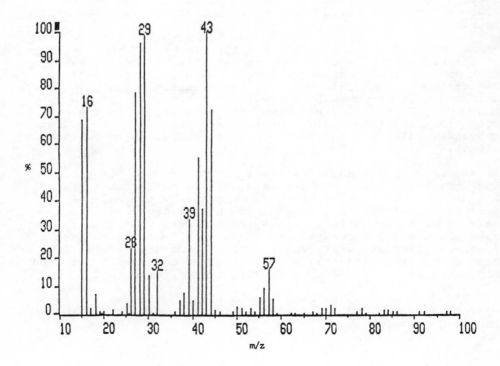

Figure 1   Direct EIMS Analysis of Gas separated from Crude Oil.

Analysis of Volatile Organic Compounds

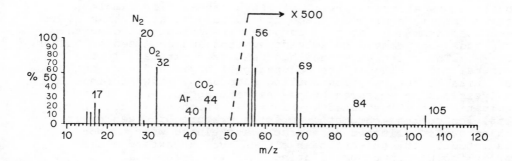

Figure 2 Direct EIMS Analysis of Contaminated Air.

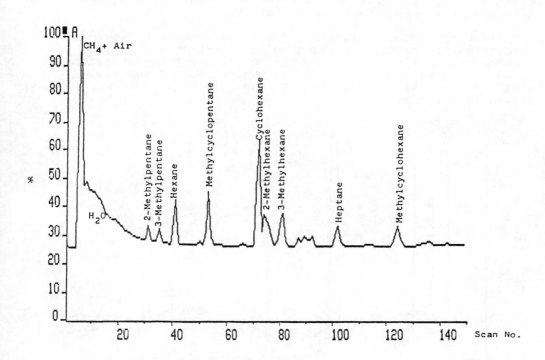

Figure 3 Direct GLC/MS Analysis of Contaminated Air (100ul gas sample)

Neither MS, GLC or GLC/MS has sufficient intrinsic sensitivity to allow the direct analysis of organic atmospheric contaminants at trace level. In such cases prior sample enrichment is required. Adsorption onto activated charcoal has been used commonly for this purpose; the concentrated organic compounds are removed by solvent elution and analysed by GLC or GLC/MS. The charcoal method has been used extensively for personal monitoring in the occupational hygiene field, but is not suitable for very low boiling compounds, which during GLC analysis can be obscured by the extracting solvent. This limitation is illustrated in Figure 4, which shows the GLC/MS total ion current chromatogram for a mixture of organic air contaminants collected on charcoal and eluted with carbon disulphide. The solvent peak elutes over an extended time period; solute compounds eluting in this region will be obscured.

This difficulty is reduced in GLC analysis since the FID response of carbon disulphide is low, but the problem remains. GLC alone, of course, lacks the specificity of GLC/MS. Peak assignment in GLC requires regular and laborious calibration of retention data with authentic reference compounds.

The activated charcoal method is unsuitable for sampling damp atmospheres, since water is retained on the adsorption tubes, affecting its retention characteristics.

The majority of the limitations discussed above are removed by concentrating organic vapours onto an adsorbent such as the porous polymer Tenax TA and analysing the resulting concentrate by thermal desorption (TD)/GLC/MS. The method is versatile, sensitive and highly specific; it also avoids the need for solvent desorption.

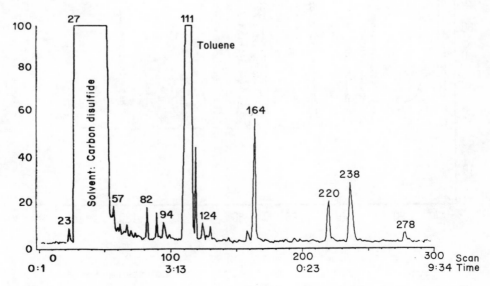

Figure 4   GLC/MS TIC Trace of Organic Air Contaminants collected on Activated Charcoal and eluted with Carbon Disulphide.

Analysis of Volatile Organic Compounds

EXAMPLES OF THE USE OF THE TD/GLC/MS METHOD

The organic vapours present at trace level in a typical laboratory atmosphere were concentrated by pumping 6l of air through a conditioned Tenax TA tube and analysed by TD/GLC/MS. The resulting total ion current chromatogram is given in Figure 5; it shows the presence of 25-30 components. These have been identified as common laboratory solvents by automatic library matching of individual mass spectra supported by manual inspection (expert validation). The assignments of the 12 major components are given in Figure 5. This example illustrates both the sensitivity and specificity of the TD/GLC/MS method. A single analysis provides a comprehensive audit of organic air contaminants and allows routine methods to be developed for the monitoring of any individual components of concern.

A similar TD/GLC/MS approach allows the detection of evaporated gasoline in a headspace sample (Figure 6). The method allows the detection of gasoline residues in, for example, soil samples following a spill. The presence of such characteristic distributions of aromatic hydrocarbons is also used forensically in the examination of fire debris during arson investigations.

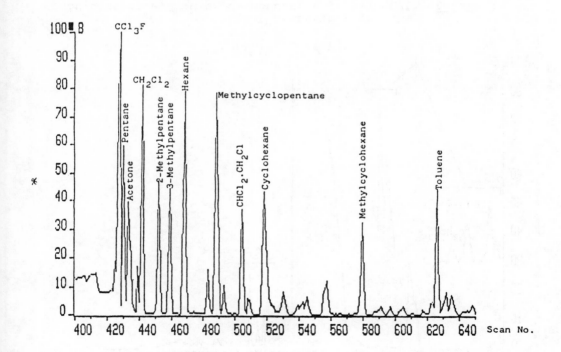

Figure 5   TD/GLC/MS Analysis of Organic Contaminants in Laboratory Air collected on Tenax TA.

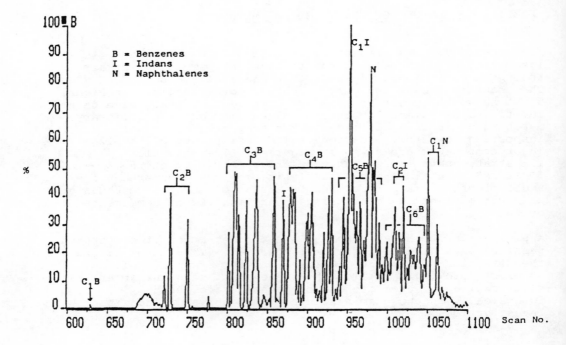

Figure 6 TD/GLC/MS of Headspace above sample containing Gasoline Residues.

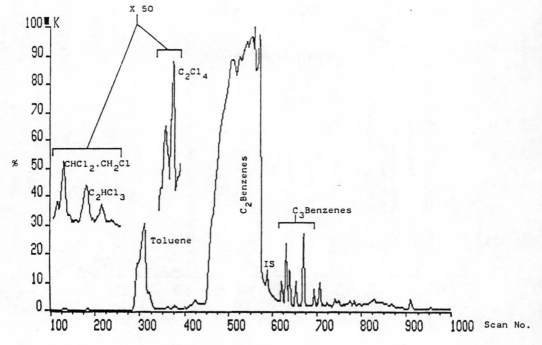

Figure 7 Organic Volatiles in a 5cm^3 Groundwater Sample Collected by Purge and Trap and Analysed by TD/GLC/MS.

Analysis of Volatile Organic Compounds

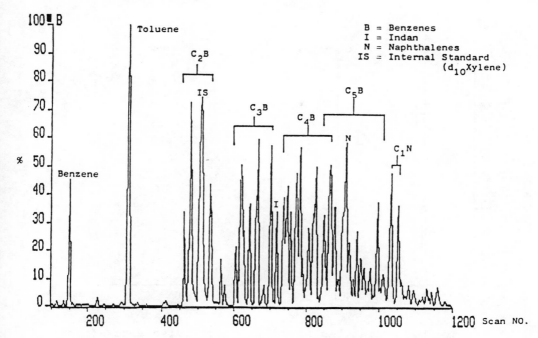

Figure 8 Organic Volatiles in a 5cm^3 Sample of Formation Water Collected by Purge and Trap onto Tenax TA and Examined by TD/GLC/MS.

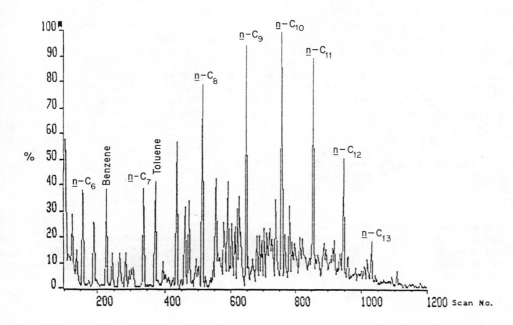

Figure 9 Volatiles in an Oily Water Discharge Before Treatment (Purge and Trap TD/GLC/MS)

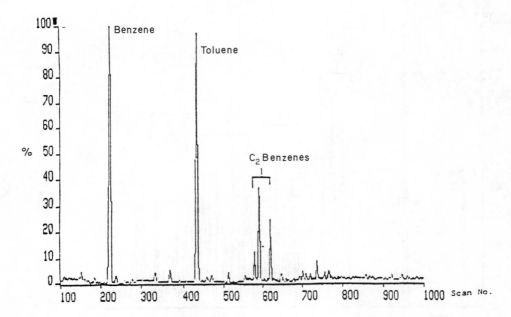

Figure 10   Volatiles in an Oily Water Discharge After Treatment (Purge and Trap TD/GLC/MS).

The TD/GLC/MS method is highly flexible and can also be applied to the specific analysis of volatile organic compounds in aqueous samples. The volatiles in question are removed from a known volume of water by purging with organic-free helium and collected on pre-conditioned Tenax TA. Quantification is performed by reference to added internal standards. A typical example is given in Figure 7; a 5ml sample of contaminated ground water was purged with helium and the concentrated organics examined by TD/GLC/MS. The sample in question was heavily contaminated with aromatic hydrocarbons. Also present were trace quantities of chlorinated solvents, which were determined in the same analysis, thus illustrating both the specificity and the wide dynamic range of the TD/GLC/MS method.

The TD/GLC/MS method is ideally suited to the specific analysis of dissolved aromatic hydrocarbons in offshore formation/production water discharge (Figure 8). Such compounds are not detected in the routine IR method for monitoring "total oil" level in such discharges. Aliphatic hydrocarbons are detectable before oil separation (Figure 9), where they are thought to be present in suspended microdroplets of oil rather than in true solution. The corresponding sample after treatment shows only aromatic hydrocarbons (Figure 10), presumed to be in true solution.

ACKNOWLEDGEMENTS

We thank Mr. M.G. Smith, Mr. A.T. Revill and Mr. H.A. Rhodes for their assistance in the above investigations and our clients for their continuing support.

Petroanalysis '87
Edited by G. B. Crump
© 1988 John Wiley & Sons Ltd

Improvements in Oil Fingerprinting: GC/HRMS of Sulphur Heterocycles

P.J.C. Tibbetts and R. Large
M-Scan Limited, Silwood Park, Sunninghill, Ascot, Berks SL5 7PZ

INTRODUCTION

Initial analysis of an oil by gas-liquid chromatography (GLC) can give only limited information concerning its source. Some oil-oil correlation studies have utilised the relative concentrations of selected isoprenoid hydrocarbons (eg pristane, I and phytane, II) for matching purposes (Brooks et al., 1969; Gearing et al., 1976; Clark and Jurs 1979). A difference between the pristane/phytane ratios of two oils can be sufficient to establish that they are not from the same source. However, if the ratios are similar, this single parameter is not sensitive enough to establish an unambiguous link between the two oils, as many oils have similar pristane/phytane ratios (Powell and McKirdy, 1973; Butt et al., 1986). To establish such a specific correlation between two crude oils (or residual fuel oils) gas chromatography-mass spectrometry (GC/MS) is necessary.

I   II

Amongst the thousands of organic compounds which have been identified in crude oils there are those known as "biological markers" (Speers and Whitehead, 1969). A biological marker is any organic compound detected in the geosphere whose basic carbon skeleton suggests an unambiguous link with a known contemporary natural product. The assemblage of biological markers in a crude oil not only reflects the environment of deposition of the original sediment but also the thermal history of the source rock (Seifert, 1978; Seifert and Moldowan, 1978). Thus their distribution is normally characteristic of a particular oil field and hence the produced crude oil (or residual fuel oil refined from the particular crude oil).

The GC/MS analysis of indigenous biological markers, particularly steranes (III) and triterpanes (IV), is now accepted as a routine oil/oil correlation method in both the environmental (eg. Dastillung and Albrecht, 1976; Barrick and Hedges, 1981; Tibbetts et al, 1982, Jones et al, 1986 and Farran et al., 1987) and geochemical fields (eg. Reed, 1977; Seifert et al., 1979; Philp, 1983 and Cornford et al., 1983). The steranes and triterpanes have highly characteristic structures and are resistant to biodegradation and weathering, being both involatile and insoluble in water (Albaiges and Albrecht, 1979). However, in areas like the North Sea where there are many geologically similar oilfields, even the sterane and triterpane fingerprints are only subtly different (eg. Bjoroy et al., 1980 and M-Scan Ltd, unpublished results) and these differences may lie within the error expected for instrumental variation.

Sterane (III)

Triterpane (IV)

where R = $(CH_2)_nH$; n = 0-2

Dibenzothiophenes (V)

In order to establish an unambiguous correlation between two closely related oils or between an oil and a source rock, it is important to compare the GC/MS fingerprints of as wide a range of compounds as possible. One class of compound that has been suggested as a marker for oil pollution is the dibenzothiophenes (V; Ogata and Miyake, 1980 and Friocourt etal., 1983). Although there is evidence that dibenzothiophenes (DBT) are affected by weathering and biodegradation after an oil is split into the environment (Ducreux et al., 1986), the rate of degradation is slow (Reed, 1977) particularly for the higher alkylated DBTs (C_2 and greater; Berthou and Vignier, 1986). Thus these compounds may prove useful for fingerprinting purposes in geochemical studies and in the case of recent oil spills.

The present paper gives examples where GC/MS examination of the DBTs has greatly facilitated the differentiation between closely related oils.

EXPERIMENTAL

Asphaltenes were removed from the oils by dilution in pentane followed by centrifugation. Aliphatic and aromatic hydrocarbon fractions were obtained by thin layer chromatography (TLC) of the pentane-soluble fraction on Silica Gel G. Bands were scraped off the TLC plate corresponding to the Rf of aliphatic and aromatic standards separately spotted on the same plate.

GC/MS analysis was carried out on a VG Analytical ZAB-HF mass spectrometer with an integrated Hewlett-Packard 5790 gas chromatograph fitted with a fused silica capillary column connected to a cold on-column injector. Sterane and triterpane fingerprints were obtained on a 25m x 0.25mm id OV-1 column with a temperature programme of 150°C to 350°C at 8°C min^{-1}. Dibenzothiophenes were analysed using a 25m x 0.32mm id SE54 column with a temperature programme of 100°C to 325°C at 8°C min^{-1}. The mass spectrometer was operated at either low resolution (ca, 1000) or high resolution (ca 5000) with a 100uA emission current and an ionisation voltage of 70eV. In the selected ion recording (SIR) mode the following ions were monitored :

Fraction	m/z	Resolution	Compound Class
Aliphatics	191	Low	Triterpanes
Aliphatics	217	Low	Steranes
Aromatics	184.0347	High	Dibenzothiophene (DBT)
Aromatics	198.0504	High	Methyl DBT
Aromatics	212.0660	High	C_2-DBT

Where only one ion was monitored continuously (single ion monitoring; SIM) the alkylated DBTs were examined using the ion m/z 197.0425, the mass chromatogram of which was recorded on a Trivector Trilab 2000 computing integrator.

RESULTS AND DISCUSSION

The m/z 217 and m/z 191 mass chromatograms of two North Sea crude oils, showing the sterane and triterpane fingerprints, respectively, are shown in Figure 1. From this data alone it would not be possible to differentiate between the two crudes. Many North Sea crudes, however, contain a characteristic triterpane thought to be trisnormoretane (TNM; Grantham et al., 1980) and in certain cases the abundance of this component relative to the C_{29} 17 alpha(H)-norhopane can lead to a differentiation between oils. However, the TNM/C_{29} hopane ratio can also be very similar in closely related oils as is the case in the present example. It was therefore considered worthwhile to examine the dibenzothiophene (DBT) distribution in the oils.

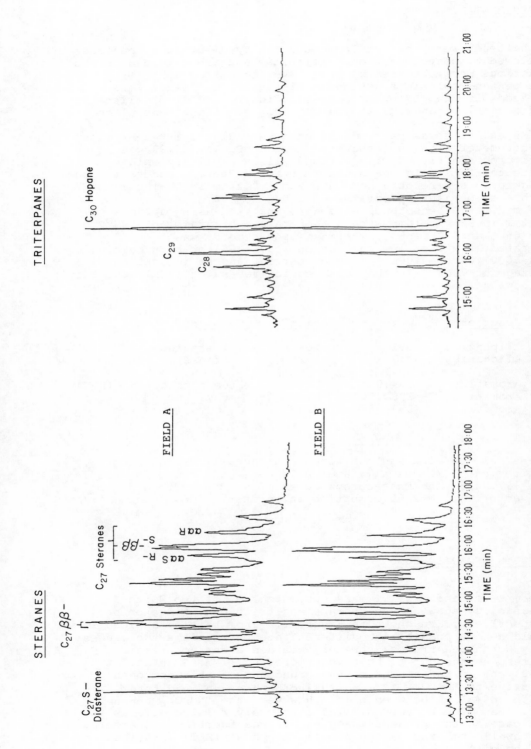

Figure 1 A Comparison of the m/z 217 (Sterane) and m/z 191 (Triterpane) Mass Chromatograms of Two North Sea Crudes.

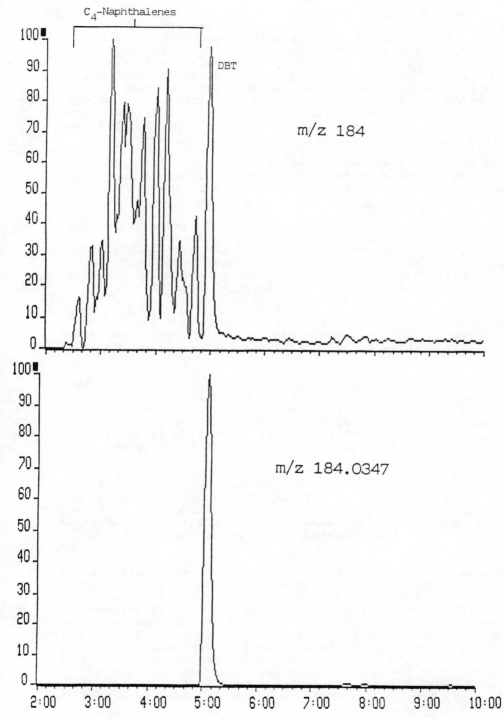

Figure 2   GLC/MS Analysis of the Dibenzothiophenes (DBT) in a Crude Oil Using Low Resolution (Top) and High Resolution (Bottom) Mass Spectrometry

With commonly used fast scanning quadrupole mass spectrometers the GC/MS correlation of sulphur-containing aromatics is complicated by overlap with isobaric aromatic hydrocarbon ions. For example dibenzothiophene and C_4 - alkylated naphthalenes have the same nominal molecular mass (m/z 184; Figure 2a). The two classes of compound can be resolved, however, by simply increasing the MS resolving power and monitoring the accurate mass of the dibenzothiophene molecular ion (m/z 184.0347; Figure 2b). The present study concentrated on dibenzothiophene itself and its methyl (m/z 198.0504) and C_2 (m/z 212.0660) homologues. The individual mass chromatogram of m/z 184.0347, being a single peak, gives no fingerprint information. However there are 4 different methyl-DBT's and a possible 20 C_2-DBT's, which produce a more complex pattern of peaks for comparative purposes.

The methyl-DBT and C_2-DBT fingerprints of the two North Sea oils discussed above are shown in Figure 3 a and b, respectively. The trace obtained for the methyl-DBTs comprises 3 peaks because the 2-methyl and 3-methyl isomers co-elute as the central peak, which has only been separated on a specially prepared capillary column (Verzele et al., 1983).

The traces in Figures 3a and b show that the two oils from North Sea fields A and B can easily be differentiated using either the methyl- or the C_2-DBT mass chromatograms.

Following the successful comparison of DBT's, a second crude oil sample was obtained from Field B. This second sample was from a different well, which was tapping oil from a separate and slightly deeper reservoir. The resulting DBT traces (Figure 3c) indicate that even these two oils from the same field can be differentiated using the DBT fingerprints.

The fingerprinting power of the GC/HRMS technique is enhanced further by combining the data for DBT and its C_1 and C_2 homologues. A comparison of a North Sea crude and one from the Middle East (Figure 4) shows a higher relative concentration of the C_1 and C_2-DBT homologues in the Middle Eastern crude.

The repeatability of the method appears good; Figure 5 shows the summed DBT mass chromatograms (m/z 184.0347 + 198.0504 + 212.0660) of the same oil analyses on consecutive days. Some differences must be expected because the mass spectrometer is scanning over three different ions. With the parameters used in the current study for selected ion recording (SIR) each ion is sampled for 70ms during a cycle time of 300ms. Therefore with capillary GC peaks only a few seconds broad it is highly probable that the top of individual GC peaks will not be sampled thus producing an error in peak height. This, of course, is true in any GC/MS technique in which more than one ion is being monitored.

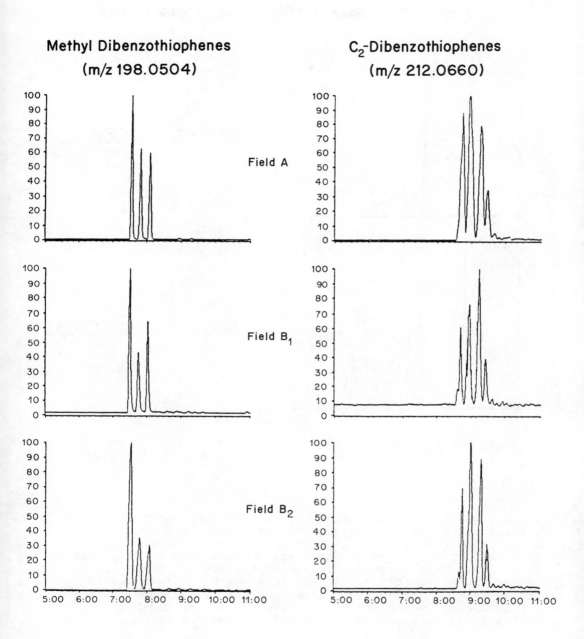

Figure 3 A Comparison of the GC/HRMS Fingerprints of the Methyldibenzothiophenes and C_2-dibenzothiophenes in three crudes from Two North Sea Fields.

(m/z 184.0347 + 198.0504 + 212.0660)

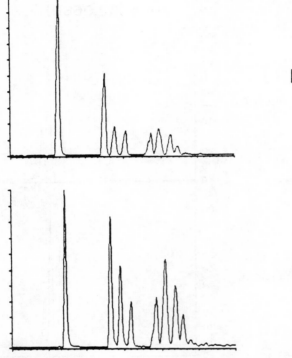

North Sea Crude

Middle Eastern Crude

Figure 4   A Comparison of the C_0-C_2 DBT GC/HRMS Fingerprints of a North Sea Crude and a Middle Eastern Crude.

In cases where two fingerprints are so similar that the differences lie within the experimental error of SIR it is advisable to resort to single ion monitoring. All of the alkylated DBT's produce strong fragment ions at a nominal mass of m/z 197. Using high resolution MS the present study monitored the ion at m/z 197.0425 in order to differentiate closely related crudes. An example of a typical trace is shown in Figure 6, in which the methyl and C_2-DBTs are clearly visible along with some minor higher homologues.

The applicability of the GC/HRMS technique has been assessed by examining fourteen different but closely related North Sea crude oil samples. Using a combination of sterane, triterpane and dibenzothiophene fingerprints it has been possible to differentiate all of the oils. The GC/HRMS technique, however, may perhaps prove to be too sensitive for some applications. For example some crudes, when transported, are a mixture of oils from several different fields (eg. Iranian Light or

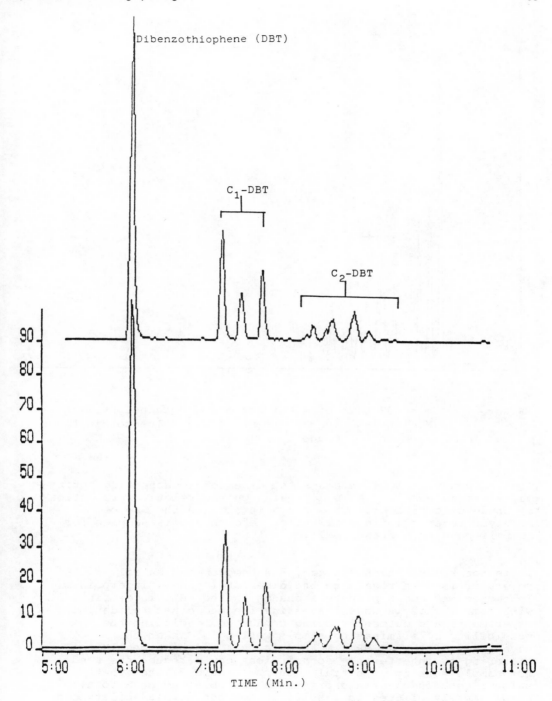

Figure 5 Two Traces obtained on consecutive days Showing the Repeatability of GC/HRMS Fingerprinting of Dibenzothiophenes.

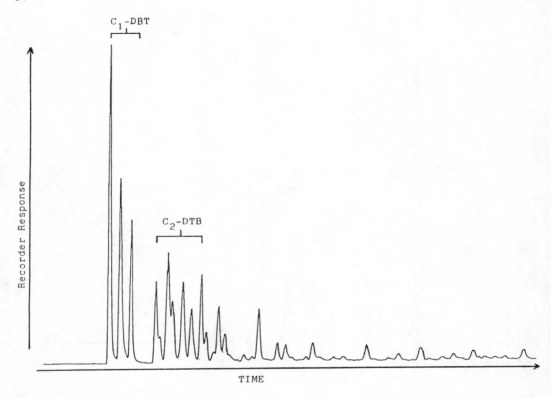

Figure 6   m/z 197.0425 Mass Chromatogram, obtained by Single Ion Monitoring, Showing the C_1 and C_2-Dibenzothiophenes in a North Sea Crude.

Iranian Heavy). In such cases the dibenzothiophene fingerprint may alter on a daily basis depending on the relative proportion of crude oils present. This could invalidate the use of archived reference fingerprints or reference oils samples for comparison with a fresh spill.

It is not known, at this stage, how homogeneous an oil reservoir is with regard to the dibenzothiophene fingerprints. Inhomogeneity could produce a change in the DBT fingerprint with time. This is under investigation and other validation experiments are currently being conducted, including the possibility of eliminating sample work-up. One of the advantages of the GC/HRMS method is that the high resolution filtering of interfering isobaric ions could reduce the need for sample work-up. Initial results (Figure 7) show that sulphur-heterocycle fingerprints can be obtained on a total oil sample merely diluted in a suitable solvent. This ability can therefore reduce the response time to a particular problem. However, total oil analysis using cold on-column injection tends to lead to rapid deterioration of the GC column resolution. Other injection techniques may therefore be necessary in order to achieve the full potential of the method.

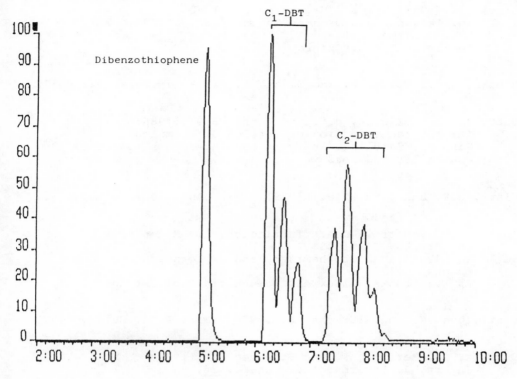

Figure 7   m/z 184.0347 + 198.0504 + 212.0660 Summed Mass Chromatogram, Showing the Dibenzothiophenes in a Total Crude Oil Sample, using GC/HRMS..

ACKNOWLEDGEMENTS

We gratefully acknowledge the support of Shell and, in particular, the useful discussions with Mr. E.R. Adlard and Mr. R. Davies. We also thank Mr. M.G. Smith for assistance with the analysis of selected oil samples.

REFERENCES

Albaiges, J. and Albrecht, P. (1979)
    Int. J. Envir. Anal. Chem., 6, 13.

Barrick, R.C. and Hedges, J.I (1981)
    Geochim. Cosmochim. Acta, 45, 381.

Berthou, F. and Vignier, V. (1986)
    Intern. J. Environ. Anal. Chem., 27, 81.

Bjoroy, M., Hall, K. and Vigran, J.O. (1980)
In: 'Advances in Organic Geochemistry, 1979' p.77
Eds. J.R. Maxwell and A.G. Douglas
Pergammon Press, Oxford.

Brooks, J.D., Gould, K. and Smith, J. (1969)   Nature, **222**, 257.

Butt, J.A., Duckworth, D.F. and Perry, S.G. (1986)
In: 'Characterisation of Spilled Oil Samples
Purpose, Sampling, Analysis and Interpretation'. p.51
John Wiley and Sons

Clark, H.A. and Jurs, P.C. (1979)   Anal. Chem., **51**, 616.

Cornford, C., Morrow, J.A., Turrington, A., Miles, J.A.
and Brooks, J. (1983).
In: "Petroleum Geochemistry and Exploration of
Europe' p.175.
Ed. J. Brooks, Blackwell Scientific Publications.

Dastillung, M. and Albrecht, P. (1976)   Mar. Poll. Bull., **7**, 13.

Ducreax, J., Berthou, F. and Bodennec, G. (1986)
Intern. J. Environ. Anal. Chem., **24**, 85

Farran, A., Grimalt, J., Albaiges, J., Batello, A.V. and
Macko, S.A. (1987).   Mar. Poll. Bull., **18**, 284.

Friocourt, M.P., Berthou, F. and Picart, D. (1983)
In: 'Chemistry and Analysis of Hydrocarbons in the
Environment' p.125
Eds. J. Albaiges, R.W. Frei and E. Merian
Gordon and Breach, London.

Gearing, P., Gearing, J.N., Lytle, T.F. and Lytle, J.S. (1976)
Geochim. Cosmochim. Acta, **40**, 1005.

Grantham, P.J., Posthuma, J. and De Groot, K. (1980)
In" 'Advances in Organic Geochemistry, 1979'. p.29
Eds. J.R. Maxwell and A.G. Douglas
Pergammon Press, Oxford.

Jones, D.M., Rowland, S.J. and Douglas, A.G. (1986)
Mar. Poll. Bull., **17**, 24.

Ogata, M. and Miyake, Y. (1980)   J. Chrom Sci., **18**, 594.

Philp, R.P. (1983) Geochim. Cosmochim. Acta, **47**, 267.

Powell, T.G. and McKirdy, D.M. (1973)
    Nature Physical Science, **243**, 37.

Reed, W.E. (1977) Geochim. Cosmochim. Acta, **41**, 237.

Seifert, W.K. (1978) Geochim. Cosmochim. Acta, **42**, 473.

Seifert, W.K. and Moldowan, J.M. (1978)
    Geochim. Cosmochim. Acta, **42**, 77.

Seifert, W.K., Moldowan, J.M. and Jones, R.W. (1979)
    In: 'Proc. 10th World Petroleum Congress, 1979"
    Vol. 2, p.423

Speers, G.C. and Whitehead, E.V. (1969)
    In: "Organic Geochemistry : Methods and Results", p.638
    Eds. G.Eglinton and M.T.J. Murphy
    Springer - Verlag, Berlin.

Tibbetts, P.J.C., Rowland, S.J., Tovey, L.L. and Large, R. (1982)
    Tox. Env. Chem., **5**, 177.

Verzele, M., Van Roelenbosh, M., Diricks, G. and Sandra, P. (1983)
    In: 'Proc. 5th Int. Symp. on Capillary Chromatography' p.94
    Ed. J. Rijks
    Elsevier, Amsterdam.

Inductively Coupled Plasma/Mass Spectrometry

A.A. van Heuzen
Koninklijke/Shell-Laboratorium, Amsterdam, (Shell Research BV), Badhuisweg 3, 1031 CM Amsterdam, The Netherlands

Inductively Coupled Plasma / Mass Spectrometry (ICP/MS) is a relatively new technique for elemental analysis based on the hybridization of a conventional ICP optical emission source and a quadrupole mass spectrometer. The use of an atmospheric plasma in conjunction with a mass spectrometer was first realized by Gray[1] in 1975. The early work was limited by serious inter-element and matrix effects, characteristic of the capillary arc (DC) plasma used. With the introduction of the ICP to overcome these problems and the sampling with apertures large enough to induce continuum flow from the bulk plasma, the basis of the commercial ICP/MS instruments was laid[2,3]. A first commercial instrument was introduced at the 1983 Pittsburgh conference. The promising properties of this new technique for multi-element analysis were immediately recognized and at the moment there are already over 100 ICP/MS instruments in operation throughout the world. A recent review of the technique has been given by Houk[4].

ICP/MS instrument and sample analysis

A diagram of the ICP/MS is given in Fig. 1. The argon plasma is induced by a high-frequency generator and has a temperature of about 8000 °C. The liquid sample is pumped into a nebulizer, which is fitted in a spray chamber. In the spray chamber the large droplets which form the major part of the sample are removed via a drain, while a small proportion enters the centre channel of the plasma torch as a finely dispersed mist. The sample is rapidly dissociated and most of the atoms are ionized. A portion of the plasma gas is extracted from the ICP into the mass spectrometer through a small sampling aperture. A molecular beam is formed in the first vacuum stage and passes through a skimmer cone into the high-vacuum stages. The latter stages contain the inon

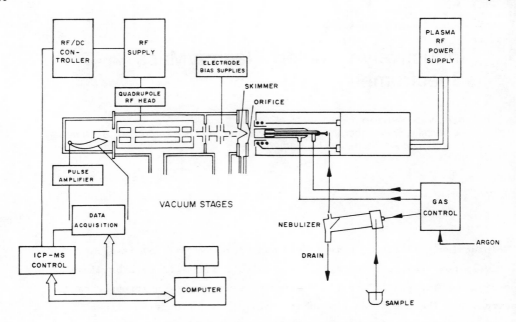

Fig. 1: Diagram of the ICP/MS system

optics and the quadrupole mass spectrometer. The ions are detected by a channeltron multiplier. Data acquisition is performed by a multi-channel analyser. The data can be transferred to a computer, where they can be manipulated and stored on disk.

Data acquisition can be performed in different modes:

- 'single ion monitoring', whereby the potentials of the quadrupole are fixed and only those ions having one specific mass/charge ratio reach the detector. In this mode the best detection limits are obtained;

- 'fast scanning', whereby a pre-selected mass range is scanned rapidly many times. In this mode a general view of the elements which are present (i.e. detectable) is obtained. This mode can be used for detecting transient signals (e.g. from electrothermal vaporization, laser ablation, chromatography) and for the determination of isotope ratios with high precision;

- 'peak hopping', whereby a number of pre-selected mass/charge ratios is chosen and the signal is detected only at those positions. This mode is useful for analytical applications where determination of a limited number of elements (with isotopes well separated on the spectrum) is necessary.

Characteristics of the technique

Low detection limits

With ICP/MS detection limits in solution are in the sub-ppb range. In general, this is one to two orders of magnitude better than with ICP/optical emission spectrometry (ICP/OES). Table 1 compares detection limits for a number of elements with those achieved by ICP/OES.

High detection limits occur for Fe and Ni. The Fe isotope with the highest abundance (mass 56, 91.7% abundance) suffers from interference from ArO^+ molecular ions. The Ni signal is distorted by small amounts of Ni originating from the sampling (and skimmer) cone.

Reference 5 indicates that it should be possible to reduce both these types of interference, and at the same time the amount of oxides (see below), by redesigning the cone.

Thus far, similar good results in terms of detection limits have been obtained for about 60 elements.

Low spectral interference

With one exception (In), all elements of the periodic system have at least one isotope with a unique mass. Combined with the simplicity of the spectra, this allows, in principle, a straightforward determination of elements at trace levels in the presence of high concentrations of other elements. If, however, the isotope with the highest abundance is subject to interference it might be necessary to use a lower-abundance isotope, which will result in a less good detection limit.

Another source of interference may be the appearance of peaks which do not originate from singly charged elemental ions. For some elements low levels of oxides and/or doubly charged ions (but only less than 1% of the singly charged ions) are found in the spectra, which may become significant if the concentration of their parent element is sufficiently high. In addition, peaks can arise from polyatomic species originating from gas and matrix molecules/elements.

Feasibility of direct introduction of organic solvents

Only small modifications to the instrument are necessary to change from aqueous to organic solutions.

Table 1

Detection limits obtained for aqueous solutions with ICP/MS and ICP/OES
and for organic solutions with ICP/MS

These detection limits were determined on a VG PlasmaQuad ICP/MS and an ARL 34000 in our laboratory. The data were obtained under 'compromise conditions' (i.e. the ICP/MS was tuned up to give more or less uniform performance over the entire mass range and the ICP/OES was set up to determine 40 elements simultaneously).

Detection limits for the organic solution were obtained with a standard solution of a 21-element metallo-organic Conostan standard (900 mg/kg), diluted to a final concentration of about 0.2 mg/l in a 1:1 (v/v) mixture of kerosine and SHELL CYCLOSOL 63.

Element	ICP/OES	ICP/MS			
	Aqueous			Aqueous	Organic
	DL^a (μg/l)	Isotope (amu)	Abundance (%)	DL^a (μg/l)	DL^a (μg/l)
Ag	1.0	109	48	0.02	0.03
Al	1.5	27	100	0.2	0.7
B	1.0	10	20	1.0	2.5
B	1.0	11	80	0.2	n.d.
Ba	2.5	138	72	0.02	0.02
Ca	2.0	43	0.1	n.d.	12
Cd	1.0	114	29	0.02	0.05
Cr	1.0	52	84	0.04	n.d.
Cr	1.0	53	10	n.d.	2.7
Cu	0.4	63	69	0.04	0.08
Fe	2.0	56	92	5.5	1.2
Mg	8	24	79	0.08	n.d.
Mg	8	25	10	0.4	6.5
Mn	0.1	55	100	0.4	0.06
Mo	6	98	24	0.05	0.05
Na	3	23	100	n.d.	0.3
Ni	15	60	26	0.4	n.d.
P	55	31	100	n.d.	2.2
Pb	26	208	52	0.02	0.08
Si	10	28	92	n.d.	n.d.
Sn	25	120	33	0.04	0.09
Ti	0.8	48	74	0.2	0.16
V	1.2	51	100	0.3	0.08
Zn	1.0	66	28	0.06	0.17

n.d. = not determined
a. DL = detection limit (2 σ) calculated from background (10 s integration)

- To prevent saturation of the vapour in the water-cooled spray chamber, the nebulizer flow has to be reduced, which results in a lower velocity of sample flow into the plasma. To obtain a comparable residence time of the analyte in the plasma to that with aqueous solutions, the aerosol injector tube of the torch is modified (ID 1.2 mm instead of 1.5 mm).

- Operating with organic solvents may cause sampling problems on account of sooting and finally blockage of the cooled sampling aperture by carbon deposits. This can be prevented by adding a small amount of oxygen via the drain to the nebulizer flow.

Detection limits obtained for a standard solution in an organic solvent are included in Table 1. Typical operating conditions, both for aqueous and for organic solutions, are given in Table 2.

Table 2

Typical ICP/MS operating conditions for aqueous and organic solutions

		Aqueous	Organic
Plasma RF power,	kW	1.35	1.35
ID of torch,	mm	1.5	1.2
Plasma gas,	l/min	14	14
Auxiliary gas,	l/min	0.5	0.7
Nebulizer flow[a],	ml/min	630	470
Oxygen flow,	ml/min	–	40
Sample uptake rate,	ml/min	1.0	1.0
Resolution quadrupole (R)[b],	amu^{-1}	1.3	1.3 (2.1)[c]

a. Meinhard nebulizer with Scott-type water-cooled spray chamber.
b. $R = 1/\Delta(M)$, where $\Delta(M)$ is the peak width at 5 % of the peak height.
c. In brackets: resolution used for Al and Ca.

A remarkable feature observed was that the organic solvents could be run at the same RF power as had been used for the aqueous solutions. This was contrary to expectation. Earlier ICP/OES experience[6] indicates that, depending on the nature of the organic solvent used, higher power is generally required to sustain the plasma. In addition, Hutton[7] has reported the use of higher power (1.8 kW) when running organics with ICP/MS.

In general, the detection limits for aqueous and organic solvents are comparable. Only for some elements with a mass below 56 atomic mass units (amu) does interference by carbon and carbon compounds occur. In these cases (with the exception of Si), however, the problem can be circumvented: either an interference-free (low-abundance) isotope line is available (B,Mg,Cr) or else higher resolution can be applied to remove interference from very high neighbour peaks, e.g. CO^+ (M=28) on Al (M=27) and CO_2^+ (M=44) on Ca (M=43). To illustrate this, the low-mass regions of a water blank, an organic solvent blank and an organic standard solution spectrum respectively are given in Fig. 2.

Yield of information about isotopes

When a mass spectrometer is used for detection, additional information about isotope ratios of an element is obtained. With the quadrupole mass spectrometer isotope ratios can be determined with high precision and these can be used, for instance, for geological fingerprinting and isotope dilution analysis.

An example of geological fingerprinting is the determination of lead isotope ratios for survey work in exploration geochemistry. Of the four Pb isotopes, one is present in a static amount due to its natural stock, while the other three also partially originate from the radioactive decay of Th and U. Most lead deposits do not contain Th and U and thus the isotope ratios of the deposits follow a 'reference curve' (often called growth curve), their positions being related to their stratigraphic age[8]. These isotope ratios are normally measured by means of mass spectrometric techniques, for which, however, time-consuming separation and pre-concentration procedures are needed. The precision of isotope ratios determined with by ICP/MS (0.1-1%) is lower than that obtained by other techniques (0.01-0.1%); on the other hand, no sample preparation steps are necessary except for partial dissolution of the Pb. Some results obtained by ICP/MS and spark source MS are given in Table 3.

Isotope dilution analysis (IDA) can be used to determine the concentration, with high precision and accuracy, of an element in a sample that has to undergo some chemical treatment before analysis. This pretreatment may lead to biased results owing to the possible occurrence of partial element loss.

Inductively Coupled Plasma/Mass Spectrometry

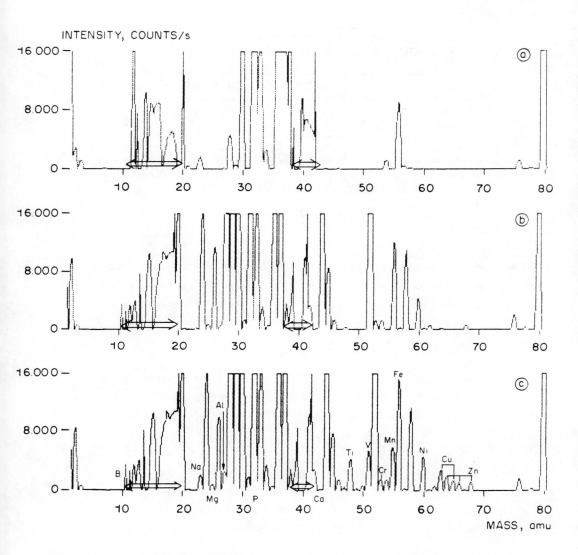

Fig. 2: Low mass region of a) water blank, b) organic solvent blank,
c) organic solvent with a 21-element Conostan standard
(concentration about 0.2 mg/l).

To prevent quick ageing of the detector the mass regions 11-20 amu (C, N
and O related peaks) and 38-42 (Ar peaks) are skipped. These regions are
indicated by ⇔; the features obtained in these regions are meaningless.

The flat tops of some of the peaks are due solely to the vertical scale
chosen.

In c) the peaks due to elements from the standard are indicated.

Table 3

Pb isotope ratios determined by ICP/MS
and spark source mass spectrometry (SSMS)

Ratio	208/206	207/206	207/204	206/204	208/204
ICP/MS precision*	2.100 0.006	0.856 0.004	15.60 0.10	18.23 0.18	38.3 0.4
SSMS	2.0969	0.8583	15.606	18.183	38.127

* The precision is given as twice the absolute repeatability standard deviation (95 % confidence level).

Basically, IDA consists of adding ('spiking') a known amount of a stable isotope of the element under study prior to the chemical treatment. In this way the isotope ratio in the sample is changed. Now, if losses occur during the subsequent treatment, all isotopes will be affected equally. The isotope ratio, therefore, will remain constant.

The actual measurement consists of determining the isotope ratio in the unspiked and the spiked sample. From the isotope ratio and the amount of spike added the original concentration in the sample can then be calculated.

An example of the use of IDA is the determination of Re in an alumina catalyst. Precision and accuracy in the determination of Re are of prime importance for the recovery of Re from the catalyst. Confirmation of a new wet chemical method for Re determination was required.

Using the correct amount of spike on the basis of prediction from theoretical principles, the overall uncertainty could be reduced to less than 1 % (at a 95 % confidence limit).

Weaker aspects of IPC/MS

Usually, after the first round of enthusiasm, the shortcomings and problem areas of a new analytical tool soon become apparent as the technique is applied to real world samples. Some of these weaker aspects of ICP/MS are the following:

- The manner of sampling from the plasma in ICP/MS restricts the concentration of dissolved solids. High concentrations will cause sooting and finally blocking of the sampling aperture. Houk[4] recommends that the total level of dissolved solids should be kept below 0.2 %. In our experience the total level depends on the elements in the solution, e.g. 1 % In solution does not give problems, while 1 % Cr solution definitely does.
- Another aspect is also concerned with a high level of dissolved solids. Surprisingly, it seems that ICP/MS is somewhat more susceptible to ionization suppression effects than ICP/OES.

  It is now known that suppression effects are strongly influenced by operating parameters. The effects reported range from signal enhancement of up to 100 % through no effect to suppression down to 50 %[9].
- As already mentioned above, polyatomic ion peaks may arise from the chemical matrix (and the gas) used. These interferences are mainly found in the region below 80 amu.

  An example of such a problem is the determination of V in a solution which contains hydrochloric acid. A $^{51}ClO^+$ molecular ion interferes with the 100% abundance peak of V (M=51 amu). Although some mathematical correction for molecular ion overlap is possible, it might sometimes be necessary to modify sample dissolution procedures to avoid the more serious interferences. Dilute nitric acid is without doubt the preferred matrix[10].

Final remarks

With all its exciting possibilities, ICP/MS is still very much in a development stage, and certainly not yet a routine technique. However, in the short time in which ICP/MS has been used, it has proved its worth in dealing with a number of different elemental analysis problems.

References
1. A.L. Gray, Analyst, 100, (1975), 289.
2. A.R. Date and A.L. Gray, Spectrochim. Acta, 38, (1983), 29.
3. D.J. Douglas, E.S.K. Quan and R.G. Smith, Spectrochim. Acta, 38, (1983), 39.
4. R.S. Houk, Anal. Chem., 58, (1986), 97A.
5. Elemental Update (VG Isotopes Ltd.), Mar 86, (1986), 1.
6. A.W. Boorn and R.F. Browner, Anal. Chem., 54, (1982), 1402.

7. R.C. Hutton, JAAS, 1, (1986), 259.
8. B.L. Gulson, G. William and K.J. Miron, Econ. Geol., 78, (1983), 1466.
9. D. Beauchemin, J.W. McLaren and S.S. Berman, Spectrochim. Acta, 42, (1987), 467.
10. S. Munro, L. Ebdon and D.J. McWeeny, JAAS, 1, (1986), 211.

Petroanalysis '87
Edited by G. B. Crump
© 1988 John Wiley & Sons Ltd

Geochemistry—More Analyses Give a Better Picture

J.R. Gray
B.P. Research Centre, Chertsey Road, Sunbury-on-Thames, Middlesex TW16 7LN

Introduction.

The BP Research Centre has seven divisions of which one, Exploration and Production (EPD), provides research and technical support to BP's Regional Exploration Offices. EPD is divided into a number of branches each specializing in a particular area of exploration or production technology.

One of these, Geochemistry Branch (GCB), is subdivided into research, service and interpretation groups. The service group, called Operations Group (OPS), performs measurements on geological samples that are direct inputs into models designed to predict :

* Where petroleum is likely to occur

* How much petroleum will be there

* And the quality of the petroleum

Thus, the more analyses we are able to make, the better our prediction will be. It is therefore a major concern of this particular group to provide an efficient low cost service.

OPS Group, has three support sections in the areas of method development, sampling techniques and automation. The automation or Systems Section, which is my concern, has in the past been involved in instrument automation projects based on micro-processor technology. For the past nine months we have been concentrating our efforts on the automation of the flow of data by developing a commercially available laboratory information management system (LIMS). This system is designed to link our laboratory instruments to a geochemistry database on another computer. This new development will form the theme of my presentation.

I intend to set the scene by describing the sort of analyses that OPS generally performs and the situation before the LIMS was developed. I will then discuss the requirements and design constraints of the system and finally take a sample's journey through the system to illustrate how it is used in practice.

The Samples.

OPS Group mainly analyses two sorts of samples :

* Sediment samples, which may be rock samples collected by a geologist in the field or well samples taken while drilling an exploration well.

* And liquid hydrocarbon samples which may be oil from a well or extract samples derived by solvent extraction of sediment samples.

Sediment samples are processed through a manual grinding, 'picking' and washing stage before being split into two. One part of the sample is decarbonated and analysed by a LECO carbon and sulphur analyser and the other part is analysed by pyrolysis and pyrolysis gas chromatography to determine organic potential and petroleum quality.

Liquid hydrocarbon samples are passed through two preparative stages, deasphaltening and HPLC analysis, which splits the original oil or extract into four fractions: asphaltenes, resins, aromatics and saturates. Geochemical parameters are derived from gas chromatography (GC) of the saturate fraction, gas chromatography mass spectrometry (GCMS) of the saturate and aromatic fractions and stable (carbon) isotope determinations of all four fractions.

Pre-LIMS Situation.

Our pre-LIMS service laboratory had the following capabilities :

* A central data acquisition system for up to 60 analogue instruments.

* Automatic data integration and data reduction methods.

* Closed loop control systems for our automatic instruments.

* Graphical data display programs.

Elsewhere on site Geochemistry Branch had a geochemical database on a VAX computer into which was manually entered data generated in our laboratory, consultants data and traded data.

Requirements and Design Constraints.

The primary requirement of the LIMS is to automate the transmission and entry of data into the VAX database in such a fashion that new items can be added 'on-line', without altering program code, as well as allowing us to target different databases planned for the future.

Secondary but important requirements were :

* A system to automatically print barcoded bottle labels specific to the sample and types of containers that were to be used for a particular sample.

* A paper free laboratory.

* To link all the instruments not just the analogue ones.

* An invoicing system.

* The incorporation of all the automatic systems that we had developed in the past.

* To make the LIMS that we developed easy to maintain.

* And, where we have to add on our own custom written code to the system, to write it in such a way that the flexibility of the original package was not impaired.

The design constraints on the system were the working methods that had evolved over a number of years and the investment of time and capital that had gone into the techniques we employ, namely :

* The pre-existing VAX database.

* The analytical instruments that we had working in the laboratory.

* The working practices used in the laboratory. The most difficult in terms of data automation being the ´casual´ way machines are used by a variety of people on unpredictable schedules with changing priorities.

* And the fact that the laboratory does not employ specialists to operate any except the most complex (GCMS) instruments.

Sample Processing.

In overview the system now consits of a sample reception station where the information relating to the identification of a batch of samples is entered, a series of tests for those particular samples are scheduled and the sample labels are printed.

This is followed by a series of optional analytical ´modules´ for acquiring the data.

Finally, when all the tests have been performed, a sample suite completion module is executed which extracts the data from the LIMS, transmits it to the VAX, prints the invoice and archives all the data and ancillary files specific to that batch.

Sample Reception Station.

The sample reception station physically consists of a computer terminal and two laser printers to print the sample container labels.

Each suite of samples we process may be one of a number of different batch types, for example sediments from a well, a suite of outcrop samples collected by field geologists or perhaps a collection of oils. Each different batch, therefore, has a different number and type of identification items that must be entered to uniquely identify the batch. Furthermore each sample within a batch may be one of a number of different sample types, for example a batch of well samples may include cuttings, cores or sidewall cores, each requiring different identification items and possibly different tests to be scheduled.

The login structure we have developed is known as a two level system, at the batch level the system asks the appropriate questions to identify that related group of samples and at the sample level the system asks the questions to identify that particular sample, distinct from the other samples and sample types in the batch, and also schedules the right tests for that sort of sample.

While these identification data are being entered for a particular batch, selected items are copied to a text file. This is accomplished by defining, in each LIMS item specification, a program called an 'evaluation routine' to be run every time data are entered into that item for a particular sample. This facility allows system designers to manipulate or verify the data if they wish to process it further than is possible with the inbuilt logical checking systems. By using an evaluation routine to transport the relevant data out of the LIMS we can combine it with other software systems without having to retype the information.

The data that are copied into this text file consists of item/data pairs that are subsequently read by the label printing software. The label production system prints labels as directed by a format file specific to that particular batch type. This approach allows us to generate a set of labels specific to the containers that the particular batch of samples will use in the laboratory: for example oil samples go through a different set of analyses than sediments and consequently require different containers with different labels. The format file consists of a series of instructions to the label generating program, these instructions include directives such as which printer to use, which character fonts to select, and a series of conceptual label sheets. Within each sheet label directives define where, how and what is to be printed on the sheet. As each label directive specifies one piece of text or LIMS item a number of these directives may be used to build up a real sample label with both text and barcoded information.

Once the labels have been printed the analyst can begin to perform analyses on the batch of samples.

Data Acquisition.

The instruments that we use to acquire data can be sub-divided into three categories :

   * Analogue instruments.

   * Serial programmable instruments, for example GCMS and MS systems.

   * And serial non-programmable instruments, or those that only perform a partial analysis.

Later I will describe later why I make a distinction between the two latter categories.

Analogue Instruments.

The analogue instruments, typically gas and liquid chromatographs, are interfaced to a central computer by a commercially available laboratory automation system. This package uses one A/D module per instrument linked to the LIMS through a digital loop communication system. These A/D modules sample the analogue data from the GC or LC at a rate of 1024 Hz, partially integrate that data, and transmit the resulting data packets back to the central computer at a selectable rate (0.5 to 32 Hz).

After each analysis the raw data are integrated either by the package or ´in-house´ application programs and are then further reduced to convert the chemical data into geochemical parameters. This system therefore is used to produce geochemical results from a physical sample.

However all the instruments in routine use must be automatic to enable them to perform a number of analyses without operator intervention in order to reduce costs. Thus we need to be able to control the instruments, we employ three control methods :

* The Laboratory Automation System incorporates control modules that operate on the loop system. These control modules provide adequate control for some functions such as auto-samplers for gas chromatographs.

* The second method we use is a 16 bit switch system which is completely independent of the laboratory automation system. It provides a cost effective method for providing simple ready signals coordinating stand-alone automatic instruments and the LIMS. It is also used for ancillary functions such as controlling chart recorder motion on automatic instruments.

* The third and most effective system is a digital communication link between the LIMS computer and a micro-processor. In this closed loop configuration the micro-processor controls an analytical instrument for the duration of one analysis, while the LIMS computer acquires the data. After each analysis the raw data are integrated and a decision is made about what the instrument will do next based on the analytical results. Instructions are then sent to the micro-processor to control its subsequent function. This configuration of micro-processor and LIMS removes the real time overhead of complex control of an analytical instrument from the LIMS computer onto the slave micro-processor and allows the functions of the analytical instrument to be monitored in a much more rigorous manner than would otherwise be feasible.

By combining the control and the data acquisition elements I have described, we have come up with concept of an analytical module. Each module consists upto three components :

* Pre-analysis information such as sample and file names, and data such as mass.

* An instrument with a closed loop control system.

* And a post-analysis data reduction scheme where, from the raw or

integrated analytical data, we can produce a reduced data report and enter the results into the LIMS.

We prepare our pre-analytical data using a data entry ´station´ which consists of a file editor that takes sample names from the barcodes, automatically generates data file names, in which is encoded both the analysis type and sample name, and uses an ´eavesdropping´ balance interface, developed in-house, to enter the mass of the sample.

Once all that information has been entered, the system has all the information required to perform a series of analyses on an automatic instrument.

After each analysis instrument specific software will execute to calculate the required parameters. Input to these applications is typically the raw data just acquired, possibly a blank run stored in a ´background´ file with which to perform a background subtraction, integrated peak areas and calibrations. It´s output will consist of the geochemical parameters, which are actually written into a user area of the data file, a reduced data report, updating the LIMS and ancillary functions such as updating a standard log file.

One of the overall aims of the system is to keep the routine day to day running as simple and as independent of computer programmers or experts as possible. One problem area in that respect is the multi-calibration system we use on some of our pyrolysis instruments. In order to get round this problem, we have developed some software that will read and evaluate formulas encoded as text in a computer file each with a specific application range. This software allows users to recalibrate instruments without reprogramming.

Digital Instruments.

The second sort of instrument classification referred to above are programmable instruments with serial communication links: for example GCMS instruments with their own computers. Computers on these sort of instruments are often supplied with non-standard operating systems or only propietry communication systems. In order to be able to pass pre-analytical data to this sort of instrument and get the results back for entry into the LIMS we have written programs on the LIMS that emulate slave versions of the propietry communication system provided with the particular instrument. This approach enables us to tackle the problem at the end where our expertise lies. Typically systems of this sort are file transports that allows the instrument operator to request pre-analytical data when he is ready to perform a series of analyses and send the results to the LIMS when he has finished.

Some of the instruments that have serial interfaces, are either not programmable or do not perform a complete analysis. One example of these is a sample preparation robotic system. The function of this system is to decarbonate sediment samples using an acid digestion technique. Samples start this process as dry rock powder, are weighed and are processed through the acid digestion and washing stages by the robot ending up as wet decarbonated samples. These are then put in an oven overnight by an operator to dry. While the samples are being processed the robotic system transmits back to the central computer the sample name, the empty tube weight, the weight of the tube plus the sample and the rack position of the sample. This information however is not the final carbonate per cent result required in the LIMS which cannot be determined until the dry samples have been weighed.

Rather than store this non-vital data in the LIMS, we prefer to use small instrument specific satellite databases. There is one such database for the decarbonation system. When the samples are dry they are removed from the oven and the dry weight information is read into the satellite database using the balance interface I described earlier. This is then combined with the wet weight by the application program, the amount of carbonate determined and entered into the LIMS.

A number of analytical modules of the three types are used to determine all the analytical and geochemical parameters required from any particular batch.

Completing the Batch.

Once all the required tests have been performed the sample suite may then be completed. At this stage the analyst may choose to schedule further tests, if a sample is of particular interest, and to deschedule unwanted test requests. The sample suite will be validated by his supervisor and then he may execute the ´logout´ module. This a four component module, consisting of :

* Data extraction from the LIMS

* Invoice printing

* The transmission of the results to the VAX

* And the archiving of all the data and ancillary files associated with the particular batch.

The first step of the logout module, extracts the data from the database after verifying that all the samples in the batch have been validated by the supervisor. The extracted data are put into a text file in the form item/data pairs.

The second step prints the invoice in three discrete phases. The application reads the price list, totals the appropriate items on the price list based on the data extracted from the LIMS and then prints an invoice according to a format file specific to the appropriate batch type. Both the price list and the format file are text files, that allow the system to be updated and maintained by non-specialists.

The price list consists of a number of items specific to one chargeable function. These items include internal codings, price and a description of the item.

The format file consists of two sorts of directives, command directives that tell the application for example which printer to use or define the current item of interest, and substitution directives that specify where on the invoice, to print information such as the description of a chargable item, its price or the total number of occurrences. Finally text entries in the format file are simply reproduced on the invoice.

The third step in the logout module is the transmission of the extracted data file to the VAX. This is achieved using a commercially available terminal emulation package that will link the laboratory computer to any remote computer using a asynchronous serial interface. This package has its own procedure language which we use to log onto the VAX through a switching system.

The file transfer is then accomplished by program-to-program communication. One record of the extracted data file is sent at a time to a receiving program on the VAX. The record format includes a 4 byte length pre-amble and 4 byte hexadecimal checksum which the receiver on the VAX decodes and acknowledges if correct. The laboratory computer program handles the retries and time-outs in case of data communication problems.

The final step in the logout module archives the data and ancillary files associated with the batch. This program extracts the file names from the data file and writes command files for two system utilities :

* Firstly the 'tape filer' is scheduled to archive all the files to a tape.

* And secondly a clone of the command interpreter, or main process, is used to delete all the files associated with the batch which are now no longer required on the system.

When this has been done all the data for that particular batch is now resident on the VAX and has been cleared from the laboratory computer.

Before the data can be entered into the appropriate VAX database it has to be reformatted and stripped of all the data not appropriate to that particular database. We have written a program to accomplish this task. This program uses the alternative eight character name (as opposed to the main sixteen character name) of each item defined in the LIMS to select the items relevant to a particular database. The first four characters of each LIMS item identifier are used to specify which database on the VAX and the second four characters specify the item identifier in that database. Thus, by defining the items correctly on the laboratory computer, and having the program on the VAX 'mask' searching the first four characters of each item in the data file, we can automatically target the data from the laboratory system to the VAX database without having to modify any software.

Summary.

In summary, with this laboratory information management system, we have linked our laboratory instruments to our VAX database and potentially to a number of VAX databases.

In the laboratory we have incorporated the serial devices as well as the analogue ones, we have developed an automatic barcode label generation system that will produce text as well as barcode labels specific to a particular batch. We have incorporated all the previously developed instrumentation control systems, and I believe retained the original flexibility of the LIMS despite enhancing its functionality with our own software.

Petroanalysis '87
Edited by G. B. Crump
© 1988 John Wiley & Sons Ltd

Environmental Monitoring of Trace Elements in Water Discharges from Oil Production Platforms

C.B. McCourt and D.M. Peers
Shell Research Centre, Thornton Research Centre, P.O. Box 1, Chester CH1 3SH

1. INTRODUCTION

Since the beginning of production from North Sea oil fields, the disposal of water from oil platforms has increased steadily. Recent estimates of the total water produced by the North Sea industry have predicted a discharge peak around 1988 - 1989 of over 100 million tonnes per year, making this effluent by far the major volume potential pollutant arising from oil operations. As a responsible operator, Shell is committed to monitoring the interaction of its operations with the environment and to minimising any adverse impact which might arise from them. Shell U.K. Exploration and Production (Shell Expro)* has therefore, for some years, co-ordinated a substantial research effort into several aspects of produced water discharge.[1] This programme has included research at a number of Shell laboratories in the areas of effluent composition, dispersion, biodegradation and toxicity. Compositional analysis has focussed on the characterisation of the effluent in terms of its organic and inorganic constituents. This paper is concerned solely with a strategy for monitoring a range of trace elements in the water discharge and as such, represents one aspect only of a much larger co-ordinated effort.

---

* Shell U.K. Exploration and Production (Shell Expro) operates in the U.K. sector of the North Sea on behalf of Shell and Esso.

2. ORIGINS OF THE WATER DISCHARGE

At the beginning of production from a North Sea oil field, the quantity of water produced with the oil is generally small, typically less than 1% of the total volume. This gradually rises as the production of water, naturally present in the reservoir in association with oil and gas, increases. This water is known as formation water. Another contributory factor to the increased production is breakthrough of sea water, which is often injected into the reservoir to maintain pressure. Produced water, i.e. the water which is recovered with oil at the production wellhead, may vary from 100% formation water to a very high content of injected water. A further contribution to discharged water arises from platforms having oil storage cells into which water is pumped (displacement water) to maintain the liquid volume. The water which is actually discharged overboard may therefore contain formation water, injection water and displacement water in a ratio which varies for different platforms in different production areas. Finally, before dumping, this water is separated from oil, using gravity separation equipment, to meet the requirements of current legislation for discharge of oil in water.

3. STEPS IN ESTABLISHING A MONITORING PROGRAMME FOR INORGANIC POLLUTANTS

Setting up a programme for monitoring the concentrations of trace inorganic pollutants in the water discharge consisted of the following distinct stages:

1. Identification of 'target' elements, likely to be a potential pollutant.
2. Selection and development of appropriate procedures for analysis of these elements in the most cost-effective manner.
3. Establishment of a sound sampling protocol.
4. Selection of locations (e.g. wells, platforms, fields) which should be monitored.

These stages are discussed individually in the following sections:

3.1 Target Elements

Since the volume of the water effluent discharged to the North Sea is obviously large, there must be systems in place to monitor for possible adverse effects on the environment. Knowledge of the composition, degradability and toxicity of any effluent is an essential first step in predicting its likely environmental impact. Even in the absence of any compositional data it might be expected that formation water, which has co-existed with oil in the reservoir at elevated temperature and pressure for geological periods of time, would differ markedly in composition to normal sea water. In terms of inorganic composition our attention focussed on metallic elements which, if present in the discharge, might represent a toxic hazard to marine life. For any element to give rise to concern it would have to:

(i) exhibit proven or suspected toxicity to marine plants and animals or human health,

(ii) be present in the water discharge at a concentration above the normal sea water background.

In the absence of any well defined criteria for controlling the discharge of inorganic components from offshore installations into the oceans, a target list of elements to be monitored was identified from pollution control guidelines for river, estuary, surface and groundwaters.[2] The guidelines divide substances into two lists. List I is sometimes called the 'black' list and List II is sometimes referred to as the 'grey' list. List I contains substances which are considered to be particularly toxic. The substances which are classed in List II are generally those which have a deleterious effect on the aquatic environment, but which are thought of as less toxic than those in List I. In addition to the elements identified in this way, any inorganic component not present in either List I or List II, but which could be detected in at least one production water sample in a significant amount, was included in the target list of elements. The final list of 21

Table I

Target elements for environmental monitoring programme

* mercury	antimony
* cadmium	molybdenum
selenium	vanadium
arsenic	thallium
zinc	titanium
copper	barium
nickel	beryllium
chromium	cobalt
lead	silver
boron	phosphorus
strontium	

* Elements identified from List I ('Black List') of WRC guidelines[2] and CONCAWE guidelines.
  All other elements identified from List II, except strontium which was present in neither list but was detected in production waters at a concentration above the normal sea level.

elements, adopted as the basis for our environmental monitoring programme, is presented in Table 1.

Although not totally comprehensive we are confident that the range of elements covered by Table 1 represents a sound platform for assessing the impact of most inorganic components which might be of environmental concern.

3.2 Analytical Strategy

It is recognised that toxicity of a given element is dependent on its chemical form, e.g. its complexed state and oxidation state. No attempt is made here to distinguish between different elemental forms, since this would be much too demanding of a routine monitoring programme. The elemental concentrations determined by the following analytical approach therefore constitute total dissolved metals, i.e. the metals in solution after filtration through a 0.45 micron filter.

The choice of techniques for monitoring the 21 target elements (Table 1) was based on the following criteria:

(i) as far as possible the technique which offers simultaneous determination of as many elements as possible, should be used,

(ii) the chosen technique should require little or no sample pretreatment,

(iii) adequate sensitivity must be achieved for each element.

The toxicity in concentration terms, of many of these elements can be defined for several types of marine organisms. Although dispersion of discharges in the vicinity of platforms is rapid, it is still desirable that the analytical technique for elements in this category is sufficiently sensitive to detect each close to the concentration that would give rise to acute toxicity in the most susceptible marine species. Adequate sensitivity in this case could be established as a target detection limit, close or equivalent to this acute toxicity threshold. For the remaining elements on the target list, where toxicity data is not immediately available, a compromise is sought between sensitivity and cost-effectiveness of the analytical approach,

(iv) where elements cannot be determined with sufficient sensitivity by the approach in (i), more specific procedures such as chemical preconcentration, followed by analysis using the same technique (or an alternative) should be applied.

Optical Emission Spectrometry (OES) using the Inductively Coupled Plasma (ICP) as the excitation source is well established in this laboratory as a rapid multielement analytical technique. A polychromator spectrometer with analytical lines available for each element in the target list was the starting point from which to develop the analytical approach. Problems in spraying solutions of high salt content into the plasma were avoided by using a cross flow nebuliser for sample introduction. This allowed the platform

Table II

Experimental details for direct analysis by ICP-OES

Spectrometer	Hilger E1000 Polychromator
Plasma	inductively coupled, argon
nebuliser	cross-flow
R.F. power	1.3KW
nebuliser flow rate	0.65 litre min^{-1}
plasma flow rate	11.5 litre min^{-1}
auxiliary flow rate	1.0 litre min^{-1}
solution uptake rate	1 ml min^{-1}

Element	Wavelength (nm)
Cr	267.7
Pb	220.3
Co	228.6
Ti	337.2
Be	313.0
B	249.7
Ba	233.5
Sr	407.8
Ag	328.0

water to be sprayed directly into the plasma. The only sample preparation required was therefore filtration through a membrane filter to remove suspended particles and oil. To minimise any spectral interferences arising from the salt content of the sample, calibration standards were matrix matched by addition of sodium chloride solution. This technique gave adequate sensitivity for nine of the twenty-one target elements (Cr, Pb, Co, Ti, Be, B, Sr, Ba, and Ag), but for the remaining elements adequate

Table III

Experimental details for analysis by ICP-OES after preconcentration procedure

Spectrometer	Hilger E1000 Polychromator
Plasma	inductively coupled, argon
nebuliser	Meinhard (concentric)
calibration standards	0.0, 0.5, 0.1 and 2.0mgl^{-1} multielement (Cd, Zn, Cu, Ni, Mo, V and Tl) in deionised water
R.F. power	1.3KW
nebuliser flow rate	0.45 litre min^{-1}
plasma flow rate	10.0 litre min^{-1}
auxiliary flow rate	1.4 litre min^{-1}
solution uptake rate	1 ml min^{-1}

Element	Wavelength (nm)
Cd	226.5
Zn	334.5
Cu	324.7
Ni	231.6
Mo	379.8
V	292.4
Tl	401.9

sensitivity could not be achieved. Experimental details are summarised in Table II.

Sensitivity for a further seven elements (Zn, Cu, Ni, Cd, V, Mo and Tl) was greatly enhanced by application of a chemical extraction-preconcentration technique before analysis by ICP-OES. In this well established technique,[3] dithiocarbamate compounds were used to remove these elements from the brine matrix by complex formation. The metals were then recovered in concentrated form in aqueous solution and analysed by ICP-OES. Experimental details for the analyses are summarised in Table III.

Table IV

Metalloids by hydride generation - AAS and mercury by cold cell AAS

spectrometer	: Perkin Elmer Model 703
hydride system	: Perkin Elmer MHS-10
reductant	: $NaBH_4$ (3%) in NaOH (1%)
purge gas	: nitrogen
light source	: electrodeless discharge lamp

Element	As	Se	Sb	Hg
Wavelength (nm)	193.7	196.0	217.6	253.6
Acid (dilutent)	1.5% HCl	1.5% HCl	3% HCl	1.5% HNO_3
Flame	air/acetylene lean, blue	air/acetylene lean, blue	air/acetylene lean, blue	no flame required

The five remaining target elements could not be determined to any satisfactory degree by the multielement atomic emission technique. Suitable, more specific procedures were therefore sought for each of these elements. Hydride generation-atomic absorption spectrometry was found to be highly effective for the metalloids (As, Sb and Se). The necessary sensitivity for mercury was achieved using cold vapour cell-atomic absorption spectrometry. An essential component of the mercury analysis was the sampling procedure (section 3.2), designed to preserve mercury in ionic form. Possible errors in analysis of the metalloids and mercury by these techniques were avoided by using a standard additions approach. Experimental details are provided in Table IV.

Finally, an atomic emission procedure for phosphorus was established, using a scanning monochromator instrument. Since the monochromator offered a choice of emission lines for this element, it was possible to select a line with improved signal-to-noise ratio and free from

Table V

Experimental details for determination of phosphorus by ICP-OES

Spectrometer	Leeman Plasma-Spec Monochromator
Plasma	inductively coupled, argon
nebuliser	cross-flow
R.F. power (KW)	1.2
nebuliser flow rate (litre min^{-1})	0.5
plasma flow rate (litre min^{-1})	12.5
auxiliary flow rate (litre min^{-1})	nil
solution uptake rate (ml min^{-1})	1.0
wavelength (nm)	214.9
calibration standards	0, 2, 5mg P litre^{-1} in 2.5% NaCl

spectral overlap problems. Details for this technique are summarised in Table V.

When the phosphorus content of the sample was below the detection limit of this technique (0.5mg litre^{-1}), a spectrophotometric procedure[4] was used to provide an improved detection limit (0.05mg litre^{-1}). The analytical approach finally adopted for each element is summarised in Table VI.

No single analytical technique was therefore identified as capable of determining each element to the sensitivity required by the monitoring programme. The range of techniques described here provided the most cost effective solution to analysis by this laboratory of key elements in the water discharge.

3.3 Sampling Protocol

Throughout our monitoring programme, careful attention was given to sampling, in particular to the nature of the containers used, to their pretreatment and to preservation of sample integrity. Two quite different sampling regimes were used depending on the analysis to be carried out.

Table VI

Techniques used for target element analysis

Element	Target detection limit (mgl^{-1}) (note 1)	Final detection limit (mgl^{-1})	Analytical technique
B	-	2	Direct ICP-OES
Sr	-	0.5	"
Cr	-	0.5	"
Pb	-	0.5	"
Ti	-	0.5	"
Ba	-	2	"
Be	-	0.02	"
Co	-	1	"
Ag	-	0.2	"
Cd	0.08	0.01	Preconcentration ICP-OES
Zn	0.17	0.02	"
Cu	0.05	0.005	"
Ni	0.3	0.005	"
Mo	-	0.05	"
V	-	0.05	"
Tl	-	0.02	"
As	3.5	0.003	Hydride generation AAS
Sb	-	0.003	"
Se	0.6	0.012	"
Hg	0.003	0.003	cold cell - AAS
P	-	0.5 (0.05)	Direct ICP-OES (spectrophotometry)

note 1 : target detection limit is based on acute toxicity thresholds, where known

Polyethylene has been found to be an excellent material for sampling a wide range of inorganic species in aqueous media, provided that the containers are suitably cleaned.[5] A washing technique involving successive treatments with dilute hydrochloric and dilute nitric acids and deionised water, was therefore used to prepare the containers for sampling. For preservation, each sample was acidified to pH <1.5 at the time of collection, by addition of nitric acid.[6] Samples taken in polyethene containers were used for analysis of all elements in the target list, with one exception. Since mercury losses from polyethene containers can occur by diffusion of the element through the plastic wall, glass containers were used for this element. Samples were preserved by addition of an oxidising agent (to prevent reduction to volatile mercury metal) and by acidification with sulphuric acid to pH <0.5. The completed sampling and analytical strategy is illustrated in Figure 1.

4. MONITORING PROGRAMME

Over 75% of the total water discharged from Shell Expro platforms during 1985 originated from four platforms, three from one producing area, one from another. Each area differed considerably in production. The monitoring program focussed on discharges from these platforms. A sampling programme was established to illustrate :

(i) differences in the composition of the discharges from different wells and from different platforms

(ii) differences in composition from the same locations over prolonged time scales.

In addition to water sampled at the production wellhead, samples were also drawn from the oil-water separator system, i.e. the water which was actually discharged to the sea. Water samples were therefore taken from several locations (wells and overboard discharge) on four platforms, throughout 1985 and 1986. In most cases samples were collected in glass (for mercury analysis) and in polyethene (for analysis of other target elements).

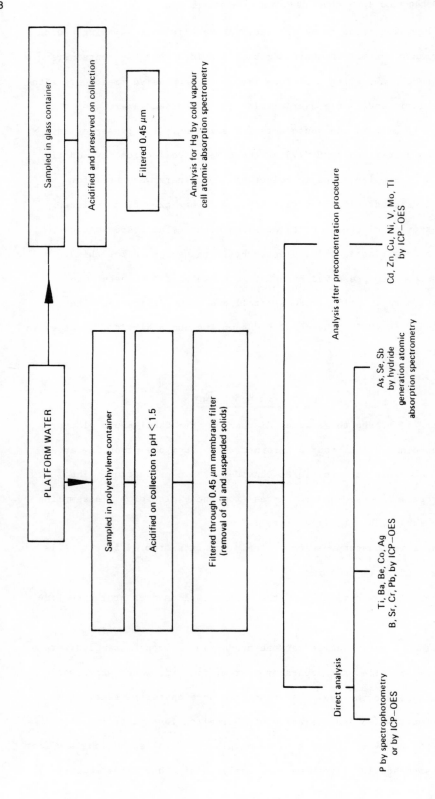

FIG. 1 — Strategy for sampling and analysis of platform waters

Table VII

Elements present in all platform water samples at concentration below detection limit of technique

Element	Concentration in all water samples (mgl^{-1})
mercury	< 0.003
cadmium	< 0.01
chromium	< 0.5
lead	< 0.5
titanium	< 0.5
beryllium	< 0.02
cobalt	< 1.0
silver	< 0.2
nickel	< 0.005
molybdenum	< 0.05
vanadium	< 0.05
thallium	< 0.02
selenium	< 0.012
arsenic	< 0.003
antimony	< 0.003

5. RESULTS

A large number of platform water samples were analysed during the monitoring exercise, leading to an extremely large amount of data. For simplicity however, results may be grouped into two categories.

(i) Elements which were found at equivalent concentration in all samples. In each case the element concentration was below the detection limit of the technique.

(ii) Elements which were found at varying concentrations across the sample population. In most cases the observed element concentrations were above the normal seawater level.

Fifteen of the twenty-one elements on the target list fall into the first category, i.e. were present in all water samples at a concentration which is below the detection limit of the appropriate technique. These elements are summarised in Table VII.

Data for the elements boron, strontium, barium and phosphorus are illustrated for a number of water samples in Figure 2 to 5. Each figure includes data from several wells on all four platforms. For simplicity no attempt is made to identify specific wells or platforms in these figures.

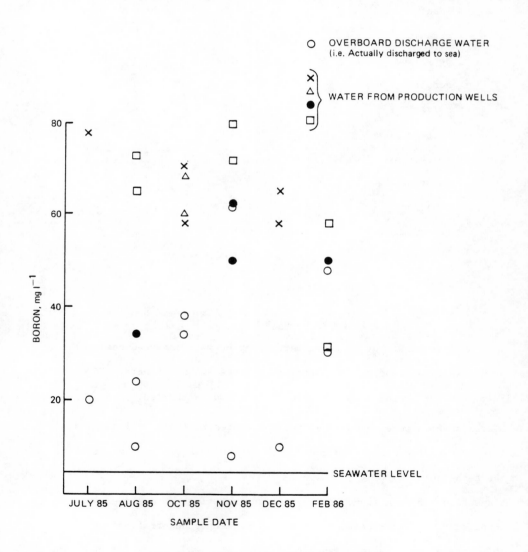

FIG. 2 — Boron in platform discharge waters

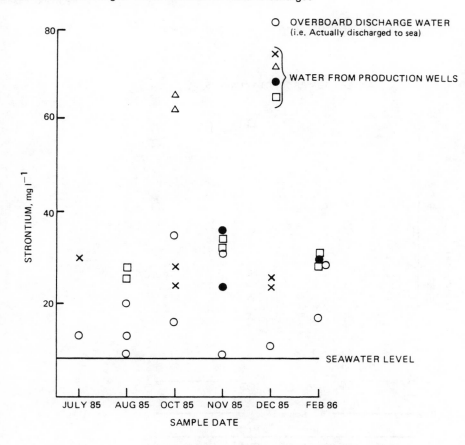

FIG. 3 — Strontium in platform discharge waters

The only exceptions to this are samples of the overboard discharge waters (the water actually discharged to the sea) which are distinguished from samples of production water.

Data for zinc and copper discharges from one platform only are illustrated in Figures 6 and 7. Similar trends were observed in discharge water from one other platform, but for all other water samples, the concentration of each metal was below the detection limit of the technique.

6. DISCUSSION

The concentrations of all of the environmentally sensitive elements in the water discharges from the four platforms are low and do not give rise

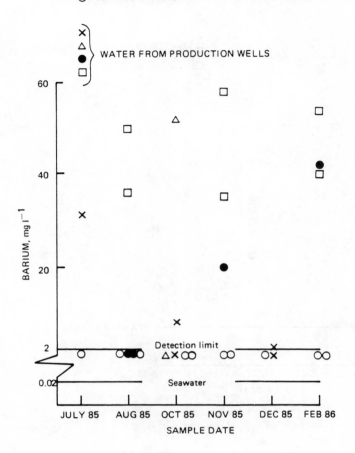

FIG. 4 — Barium in platform discharge waters

to concern. Significantly, the high toxicity elements mercury and cadmium are not detected in any discharges. A few elements are present in the discharge at concentrations slightly above the normal sea water level. An important consideration in assessing their impact is the massive diluting action of the surrounding sea water on the effluent plume. Water discharge from Shell Expo platforms totalled 22 million tonnes in 1985. For the U.K. sector the total water body is about $1.5 +10^{13}$ tonnes. Dispersion of the discharge plume is rapid and it is likely that dilution factors of several hundred-fold are achieved within a few hundred metres of the platform.[7]

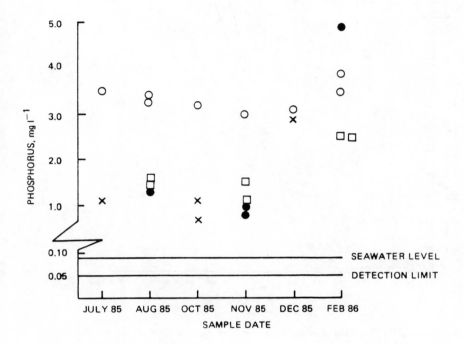

FIG. 5 – Phosphorus in platform discharge waters

Thus, although the discharges of zinc and copper (Figures 6 and 7) from some platforms were occasionally somewhat greater than normal sea water levels, there is little doubt that these were rapidly reduced by dispersion to normal background levels. The origins of the zinc and copper discharges are not clear.

The discharges of boron, strontium, barium and phosphorus, as illustrated in Figures 3 to 5, are easier to explain. Boron occurs in formation water, typically at a concentration of 60 to 80mg litre^{-1}, and is therefore invariably present to some extent in production water. Dilution by

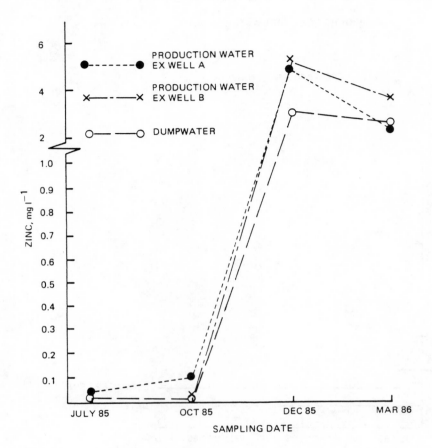

FIG. 6 — Zinc in water discharge from one platform

water from other sources, (e.g. displacement water), reduces boron in the overboard discharge to levels which are substantially less than the production water and only slightly above sea water, as indicated by Figure 3. Seawater dilution of the discharge plume easily reduces boron to the normal ocean concentration. Similarly, strontium can also occur in substantial concentration in formation water, but the eventual strontium discharge is only slightly greater than the normal sea water level (Figure 4) and is rapidly restored to normal levels by dilution of the plume.

Barium occurs in formation water, typically at concentrations of 50 to 60mg litre^{-1}, but precipitates as $BaSO_4$ on contact with sea water.

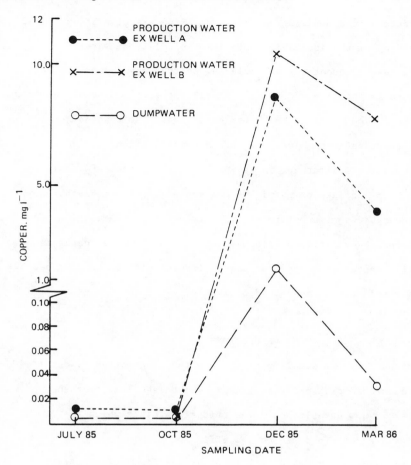

FIG. 7 — Copper in water discharge from one platform

Therefore, although barium is detected in many production water samples at concentrations well above the normal sea water level, the barium discharge in the overboard discharge water (i.e. after mixing with sea water) is always less than 2mg litre^{-1}. Indeed, barium sulphate precipitation (as scale) in the tubing of producing wells is a phenomenon which can reduce productivity. Scale formation is therefore inhibited by treatment of the injection water with phosphonates and/or other chemicals and these phosphonates are almost certainly the source of the phosphorus levels observed in the discharges. As with other elements, the phosphorus discharge is rapidly diluted to normal sea water levels.

The data accumulated during this monitoring exercise has provided a sound understanding of the inorganic composition of platform water. Monitoring on a less frequent basis continues, so that any change in the established trend is detected. Future work will focus on some other aspects of trace inorganic components in the water discharge. The tendency of metals, which are present in solution at trace concentration, to adhere to the surface of particulate matter, is well known. The extent of metal accumulations on the surface of particles present in the water discharge will therefore be the subject of a future study. Trace metal accumulations in mussels taken from the vicinity of discharge areas will also be established. Current work will assess the impact of production activity on the sea bed by study of trace metal levels in sea bed sediments.

7. CONCLUSIONS

A methodology for sampling platform waters, supported by detailed analytical procedures and associated data on analyte sensitivities and detection limits is now established. The criteria against which the detection limits are set are based on the efficient utilisation of multi-element techniques supported by more sensitive element-specific methods, wherever these are needed, to improve detection limits.

Direct multi-element analysis by ICP-OES provides satisfactory detection limits for the elements Cr, Pb, B, Sr, Ti, Ba, Be, Co and Ag. Target sensitivities are achieved for the elements Zn, Cd, Cu, Ni, Mo, V and Tl by a preconcentration-extraction procedure (using APDC-DDDC) prior to analysis by ICP-OES. Optimum sensitivities are achieved for elements of the metalloid group, using hydride generation-atomic absorption spectrometry and for mercury using cold vapour cell-AAS. Phosphorus is determined independently by sequential ICP-OES.

An extensive data base is now established on which to base the assessment of environmentally sensitive elements in water discharges from Shell Expro oil production and drilling platforms. The concentrations of these elements in the water discharges are low. In particular, they are below the levels which would present a hazard to the environment. Notably the concentrations of the 'black-listed' elements mercury and cadmium are substantially below the levels which are known to produce acute toxicity in marine species.

With the exception of a few elements (i.e. zinc, copper, boron, phosphorus and strontium) the inorganic composition of the discharges from different platforms is broadly similar. For most elements significant variations with time and location are not observed.

ACKNOWLEDGEMENTS

The authors gratefully acknowledge the contributions of the following (present and former) colleagues to this work:
Dr P.G. Gadd, Mr S.W. Clark, Mr G. Williams and Mr S.T. Holding, and to Environmental Affairs Department, Shell Expro, Aberdeen, for their help and assistance.

REFERENCES

1. Somerville, H.J. (1984) North Sea Oil Exploration and Production: Interactions with the Environment, in "Offshore U.K. 1984". Proceedings, Institute of Petroleum Annual Conference, 100-121.
2. J. Gardiner and G. Mance, July 1984. Water Research Centre Technical Report TR204.
3. McCleod, C.W. et al., Analyst, April 1981, 106, 419.
4. Murphy and Riley, Anal. Chim. Acta, 27, 31 (1962)
5. Moody, J.R. and Lindstrom, R.M, Anal. Chem. $\underline{49}$, 2264, (1977).
6. CONCAWE Report No. 6/82.
7. Somerville, H.J., et al, Marine Pollution Bulletin, 1987, (in press).

FABMS Analysis of Surfactants and Polar Petroleum Compounds

R. Large, P.J.C. Tibbetts and A.J. Holland
M-Scan Limited, Silwood Park, Sunninghill, Ascot, Berks SL5 7PZ

INTRODUCTION

Fast atom bombardment mass spectrometry (FABMS) was developed initially for the examination of biopolymers, such as peptides and carbohydrates, which are difficult to analyse by other MS methods (Barber et al (1981); Morris et al (1981)). The FABMS technique, however, is also proving useful for the analysis of surfactants (anionic, cationic and non-ionic) and other polar petroleum-based compounds (Tibbetts et al (1985); Holland et al (1986); Bare and Read (1987); Lyon et al (1984)).

FABMS is essentially a "soft" ionisation method, which provides specific molecular weight information for otherwise intractable or labile samples, but in limited cases structural information can be deduced from the appearance of weak fragment ions; for example small peptides can be sequenced fully on the basis of a FAB spectrum. In these respects the FABMS technique complements the more conventional MS ionisation methods (electron impact/chemical ionisation (EI/CI)) and has largely superseded the experimentally difficult field desorption (FD) technique. The FABMS method is particularly powerful when used in conjunction with a high field mass spectrometer; FAB spectra have been recorded for biochemical compounds of molecular weight up to 20,000 daltons (Bell and Green (1987)). High resolution FABMS can allow molecular formulae to be determined unequivocally.

The particular advantage of the FABMS method is that samples are examined in the solution phase; unlike EI/CI there is no requirement to vaporise the sample before MS ionisation. High boiling and/or thermally labile compounds may therefore be examined directly. Typically a solution of the sample in question is added to a bead of glycerol on the metal FAB target; evaporation of the solvent leaves the solute dispersed through the glycerol matrix. Solute molecules diffuse to the surface of the matrix where they are ionised by bombardment with highly energetic inert gas atoms and/or ions; xenon is normally used for this purpose. FABMS ionisation proceeds by addition of a proton or cation to produce positive ions and/or

by the loss of a proton to produce negative ions. Acidic
petrochemicals (phenols, carboxylic acids) deprotonate to
anions; certain base/neutral petrochemicals (amines, amides,
ethoxylates) cationise to give positive ions. Anionic
surfactants (ether sulphates, alkylbenzene sulphonates) are
detected as intact ions, as are cationic surfactants and other
alkyl ammonium/quaternary species (e.g. biocides).

The FABMS method has been applied extensively by ourselves to
the analysis of oils and aqueous effluents, particularly those
from oil producing, storage and refining facilities. The FABMS
method is normally applied to an acidic or polar fraction
separated from the sample in question. Certain classes of
compound, such as hydrocarbons, are effectively transparent to
the FABMS method. The technique may therefore be applied to
the direct analysis of unextracted samples. In this respect
FABMS is proving to be a useful and cost-effective analytical
screening method.

EXAMPLES

The particular power of the FABMS method may be illustrated by
reference to a number of specific industrial and environmental
problems, which have been solved in a simple and elegant manner
with FABMS.

Surfactants

The FABMS method is particularly well suited to the analysis of
surfactants, presumably because such compounds disperse
uniformly within the FAB matrix; typical examples are given in
Figures 1 and 2.

The non-ionic surfactant nonylphenol ethoxylate (I; 6.5 mole)

C_9H_{19}—⟨aryl⟩—$(OCH_2CH_2)_nOH$

$C_nH_{2n+1}(OCH_2CH_2)_mOH$

I   II

shows two clear cation series under positive ion FABMS
conditions (Figure 1):

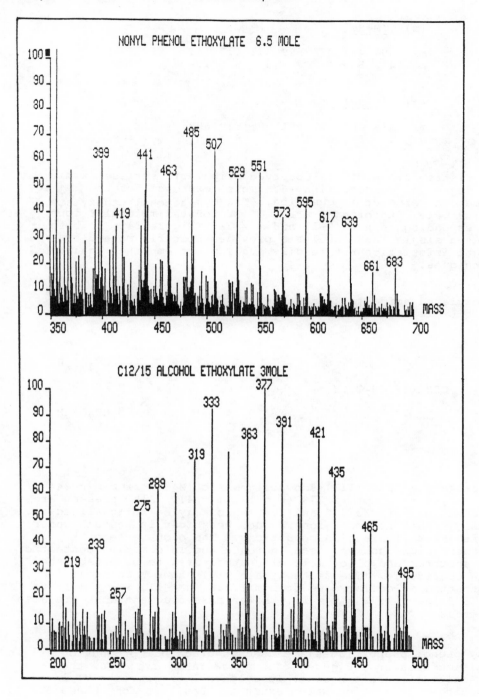

Figure 1   Positive Ion FAB Mass Spectra of Non-ionic Surfactants

(a) $(M+H)^+$

m/z	397	441	485	529	573	617	661
n	4	5	6	7	8	9	10

(b) $(M+Na)^+$

m/z	419	463	507	551	595	639	683
n	4	5	6	7	8	9	10

The distributions of the observed oligomeric series maximise at the degree of condensation (6.5 mole) expected for the commercial product in question. C_{12}/C_{15} alcohol ethoxylate (II; 3 mole) exhibits an array of protonated molecular ions corresponding to n = 12-15 and m = 2-5 (Figure 1). In both cases a single FABMS analysis provides specific molecular weight information not readily available from other analytical techniques.

Ethoxylated amide (alkanolamide) surfactants (III, IV) also exhibit clear protonated molecular ions (Figure 2).

$$C_nH_{2n+1}CONH(CH_2CH_2O)_mH$$
III

$$C_nH_{2n+1}CON\begin{matrix}(CH_2CH_2O)_xH \\ (CH_2CH_2O)_yH\end{matrix}$$
IV

Monoalkanolamides (III) show oligomeric $(M+H)^+$ from m/z 244 (n = 11, m = 1) to m/z 376 (n = 11, m = 4). Also present are intense ions of m/z 226 and 254, which may be rationalised as highly stable fragment ions. Such behaviour is unusual under "soft" FABMS ionisation conditions. The occurrence of fragmentation, however, can increase the information content of the spectrum and allow molecular structures to be established with confidence. The diethoxylate (IV) shows homologous protonated molecular ions from m/z 288 (n = 11, x = y = 1) to m/z 372 (n = 17; x = y = 1). Lower mass fragment ions are also apparent.

Alkylphenol propoxylates are used in crude oil production as emulsion breakers; Figure 3 shows how FABMS can allow the detection of such a product (octylphenol propoxylate V) in a discharged production water extract. Positive ion FABMS of

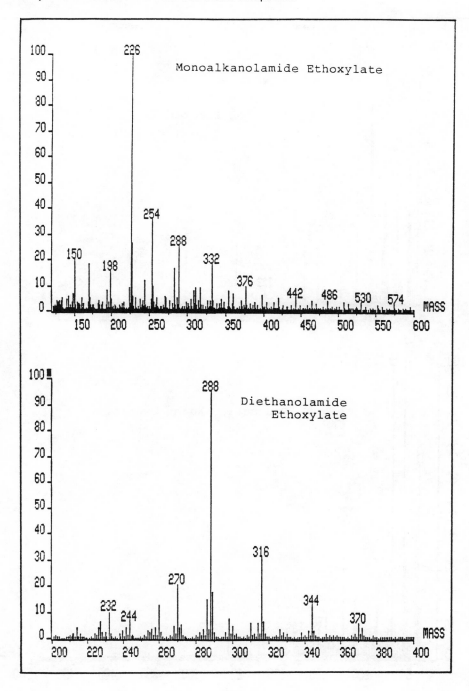

Figure 2   Positive Ion FAB Mass Spectra of Alkanolamide Detergents.

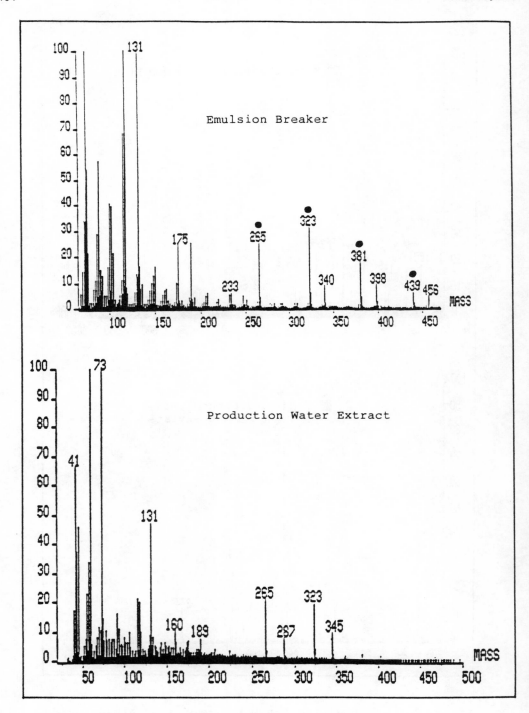

Figure 3   Positive Ion FABMS Detection of an Emulsion Breaker in a Produced Water.

FABMS Analysis of Surfactants and Polar Petroleum Compounds

C_8H_{17} — [benzene ring] — $O(CHCH_2O)_nH$ with CH_3 branch

V

the product shows protonated molecular ions corresponding to n = 1,2,3 and 4 (m/z 265, 323, 381, 439). The mono- and di-propoxylated compounds can be clearly seen in the production water extract as $(M+H)^+$ and $(M+Na)^+$ FABMS ions.

Anionic surfactants, such as alkylbenzene sulphonates, are best examined under negative ion FABMS conditions. Figure 4 shows a typical example. The molecular weight distribution of the commercial product can be established from the distribution of homologous sulphonate anions. Also present in Figure 4 are ions arising from the glycerol matrix used (m/z 91, 183); this is a typical feature of FAB mass spectra.

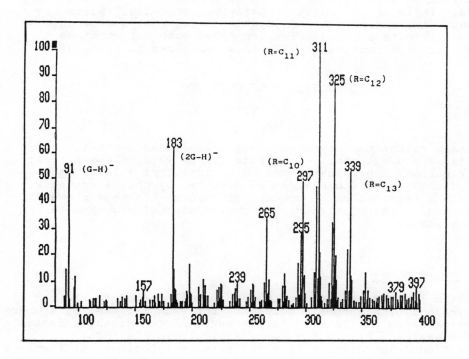

Figure 4 Negative Ion FAB Mass Spectrum of a Commercial Blend of Sodium Alkylbenzene Sulphonates ($RC_6H_4SO_3Na$)

FABMS is a solution technique. Particular solute species can be enhanced by doping the matrix with particular additives. Such a procedure allows the detection of sodium petroleum sulphonates (VI) in mineral oils (Figure 5). In the absence of matrix additives no FABMS ions were detectable. At each carbon

$$C_nH_{2n+z}SO_3Na$$

VI

typical ions	z	n	hydrocarbon compound type
353,367,381	−7	20,21,22	alkylbenzene
351,365,379	−9	20,21,22	tetralin
349,363,377	−11	20,21,22	bicyclicnaphthenobenzene
347,361,375	−13	20,21,22	tricyclicnaphthenobenzene

number (Figure 5), four separate sulphonate species are evident; these are rationalised above.

A particular advantage of the FABMS technique is that, in many cases, little or no sample preparation is needed. Surfactants, for example, can be detected by the direct analysis of aqueous samples (Figure 6). Negative ion FABMS examination of an aliphatic ether sulphate (VII) in water shows an array of

$$C_mH_{2m+1}(OCH_2CH_2)_nOSO_3^-$$

VII

sulphate anions corresponding to $m = 12\text{-}15$ and $n = 0\text{-}5$. Similarly a nonylphenol ethoxylate (I) in water shows a clear distribution of protonated molecular ions (m/z 353-749; $n = 3\text{-}12$). Related compounds can be detected by the direct negative ion FABMS examination of aqueous effluents; Figure 7 shows the detection of phenol, cresol and cyclic alcohol propoxylates in such a sample.

Naphthenic Acids

Negative ion FAB mass spectrometry is well suited to the examination of naphthenic acids in, for example, fuel oils where their presence at elevated levels may lead to extensive damage to the engine. Figure 8 shows a typical example. The conventional approach to this analysis is laborious; it involves extraction of acidic components from the oil sample, derivatisation of the extracted organic acids and analysis by GLC/MS. The resulting total ion current trace (Figure 8) shows a broad unresolved complex mixture comprising the trimethylsilylesters of the very many possible isomeric and

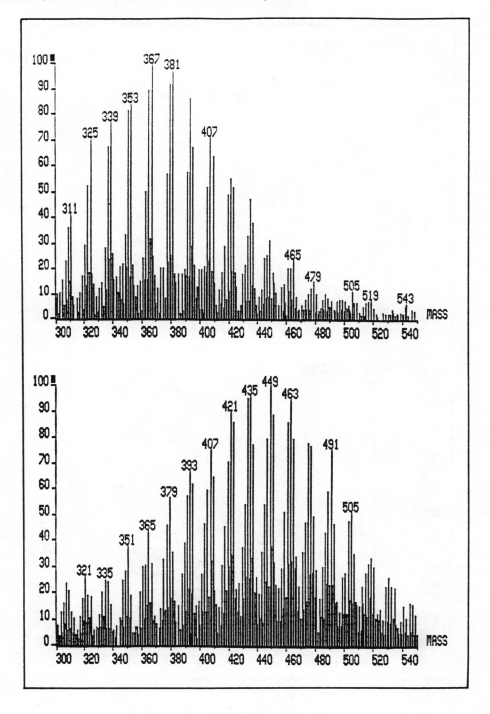

Figure 5   Partial Negative Ion FAB Mass Spectra of Sodium Petroleum Sulphonates in Oil

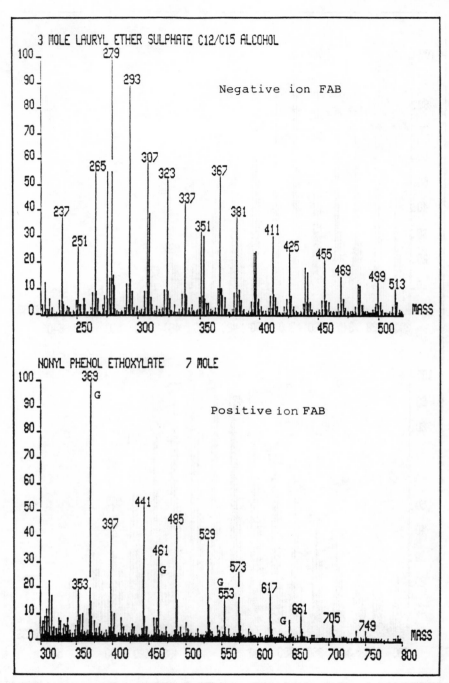

Figure 6   Direct FABMS Analysis of Surfactants in Water

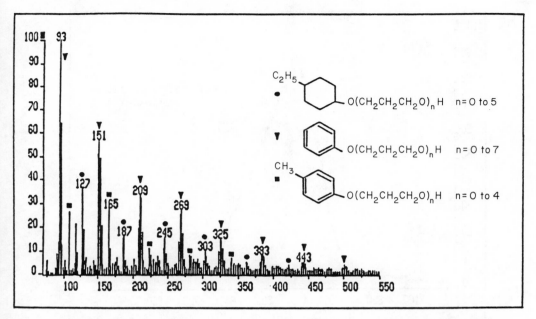

Figure 7 Direct Negative Ion FABMS Detection of Phenol, Cresol and Cyclic Alcohol Propoxylates in an Aqueous Discharge

homologous naphthenic acids. Direct negative ion FABMS analysis of the acid extract, however, shows a very much simpler picture with two principal anionic series. These correspond to carboxylate ions from bicyclic and tricyclic naphthenic acids :

	VIII			IX	
R:	C_2	C_9		C_0	C_4
m/z	209	307		235	291

Detailed inspection of the GLC/MS data allows confirmation of the FABMS assignments.

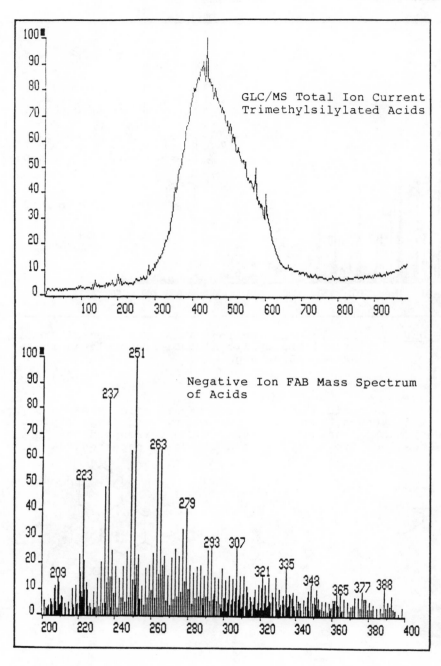

Figure 8   Analysis of Naphthenic Acids in a Fuel Oil

Petrochemicals

The use of FABMS to monitor polar petrochemicals has been reported (Tibbetts et al, 1985). Detailed and time consuming GLC/MS analysis of the derivatised acid extract of the aqueous effluent from a petrochemical site showed a number of phenols and phenol/formaldehyde/furfural condensation products (Figure 9). Each effluent component contains an acidic (phenolic) proton, which is readily detached under negative ion FABMS conditions. The resulting FAB spectrum (Figure 9) shows an $(M-H)^-$ ion for each of the components detected by GLC/MS. FABMS also reveals compounds not apparent in the GLC/MS analysis and confirms the molecular weight of certain compounds which show no molecular ions under GLC/MS conditions.

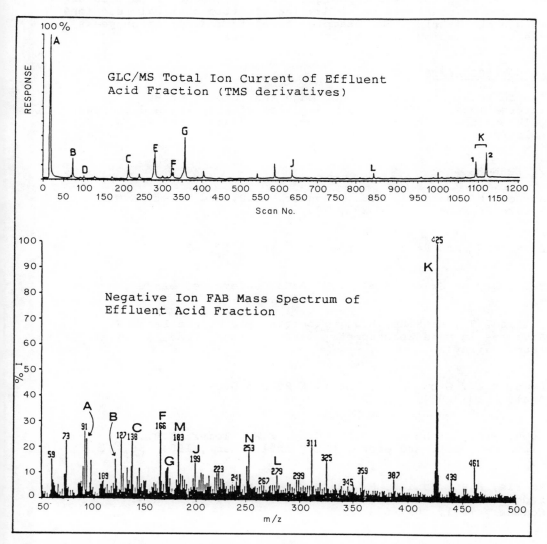

Figure 9   GLC/MS and FABMS Analysis of an Aqueous Petrochemical Effluent

This example illustrates the power of FABMS as an environmental screening method. Relatively large numbers of samples may be examined rapidly with minimal work-up to map out a zone of effect (if any) from a particular point discharge. Localised contamination can than be investigated in detail by conventional extraction/GLC/MS.

The specificity of the FABMS method may be increased significantly by determining accurate molecular masses under high resolution conditions. This enables molecular formulae of unknown components to be determined unequivocally and is particularly useful in cases, as above, where certain compounds are not amenable to GLC/MS analysis.

As FABMS is a solution technique, the assignment of particular ions can be confirmed by conducting specific chemical reactions within the matrix. For example the assignment of amine peaks can be checked by acetylation or deuteration.

ACKNOWLEDGEMENTS

The authors gratefully acknowledge the support of BP International, FOBAS Lloyd's Register of Shipping and Buchanan Oils and the contributions of Dr. N.J. Haskins and Mr. J. Holland.

REFERENCES

Barber, M., Bordoli, R.S., Sedgwick, R.D. and Tyler, A.N., Nature (London), 1981, 293, 270.

Bare, K.J. and Read, H., Analyst, 1987, 112, 433

Bell, D. and Green, B.N., 1987 ASMS Conference, Denver.

Holland, J., Haskins, N.J., Tibbetts, P.J.C. and Large R., Advances in Mass Spectrometry 1985, (Editor: Todd, J.F.), 1986, (Wiley)

Lyon, P.A., Crow, F.W., Tomer, K.B. and Gross, M.L., Anal. Chem 1984, 56, 2278.

Lyon, P.A., Stebbings, W.L., Crow, F.W., Tomer, K.B., Lippstreu, D.L. and Gross, M.L., Anal. Chem., 1984, 56, 8.

Morris, H.R., Panico, M., Barber, M., Bordoli, R.S., Sedgwick, R.D. and Tyler, A.N., Biochem. Biophys. Res. Commun., 1981, 101, 623

Tibbetts, P.J.C., Haskins, N.J., Holland, J. and Large R., Intern. J. Environ. Anal. Chem., 1985, 21, 199.

POSSUM—A Method for the Determination of Light Ends in Unstabilised Crude Oils and Condensates

H. Fitzgerald
Moore, Barrett & Redwood Ltd, Rosscliffe Road, Ellesmere Port, Cheshire

INTRODUCTION

The determination of light ends, nitrogen, carbon dioxide and methane through to pentane (C1 to C5), in pressurised crude oils and condensates is increasingly required as more shared pipelines are operated, since the value of each partner's oil contribution normally depends, amongst other factors, on the light ends composition of the crude.

The traditional way to carry out this analysis is to flash separate the oil into stable liquid and gas fractions, under controlled conditions of temperature and pressure. The two fractions are then analysed and the composition of the original fluid found by recombining the two analyses on a mass basis using the gas to oil ratio determined during the flash procedure. In the UK, IP methods 344/80 and 345/80 are normally employed for the oil and gas samples.

Some systems carry out the flash stabilisation offshore, whilst others return samples of unstabilised fluid to an onshore laboratory for stabilisation. In both cases, the costs of the dedicated helicopter flights needed to carry these samples are considerable. Offshore stabilisation requires large and complex systems to be installed on the production platform, whilst onshore stabilisation needs expensive single phase sample cylinders.

In 1980 it seemed feasible that unstabilised crude oil could be injected directly into a gas chromatograph, and the pressurised oil syringe sampling method (Possum) was developed to achieve this. We opted to use an external standard method, weighing the injection into the GC which had been previously calibrated in counts per microgram of component, whilst at about the same time another group (refs 1&2) were working on direct injection of crudes using an internal standard method based on IP 344/80 "Light hydrocarbons in stabilised crude oils by gas chromatography".

In the Possum method, the sample is drawn from the oil flow line into a syringe and then injected directly into a gas chromatograph. There are no intermediate sample handling steps or sample preparation steps required and hence no subsampling with its attendant dangers of non-representivity.

METHOD

Syringe

The syringe used is a 100 µl Series B manufactured by Precision Sampling of Baton Rouge, Louisiana, and is sold in the UK under the trade name 'Pressure Lok'. It is fitted with a twist lock valve behind the needle and a plunger blow-out stop. (figure 1)

The manufacturers guarantee this syringe to be leak tight to 250 psi (17 bar). It has been found however, that provided the syringe has been correctly pressure tested it can be used safely at much higher pressures. A number have been pressure tested to 75 bar, and a few tested to destruction. When failure does occur, it is normally in the form of small cracks in the glass barrel which instantly relieve the pressure in such a small volume. Whilst this obviously results in the sample being destroyed, no catastrophic failures which could cause injuries have been observed. The syringes have regularly been used for about seven years to sample crudes at 30 bar, and many syringes have completed more than 200 round trips from the sample point to the lab.

The series B syringe with a removable needle has proved to be more convenient than the series A with a fixed needle, and is also much easier to clean.

The dead volume in the valve should be reduced in new syringes by removing the teflon packing and cutting about 2 mm off its length.

Since the teflon plunger tips tend to "cold creep" whilst in storage, all new syringes and any syringe which has not been recently used should be treated by holding the plunger in boiling water for about 5 minutes, allowing it to cool, and then gently rubbing the tip on a polished surface to flatten and broaden it.

The front packing in the syringe should be adjusted when new, and then at periodic intervals to ensure that the valve open-close marks are aligned and that the syringe fills easily.

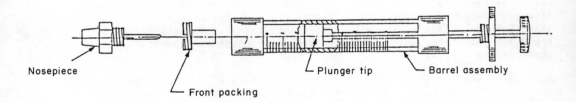

FIGURE 1   DIAGRAM OF PRESSURE-LOK SYRINGE

Sampling

The sample is drawn from the pipeline or vessel through a septum fitting which is sealed with a standard HPLC triple layer septum, see Figure 2.

It is obviously essential that the oil which is drawn into the syringe is a representative sample, and it is important therefore that it is not

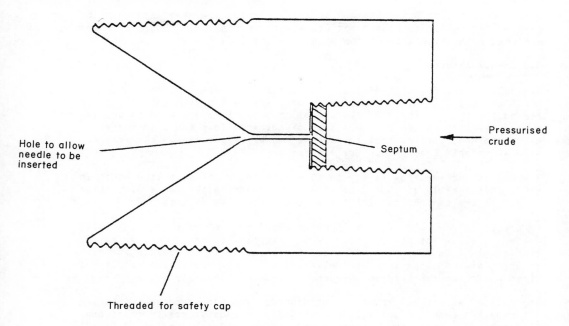

FIGURE 2   SEPTUM FITTING

subjected to any pressure drop which might cause gas break-out and subsequent compositional changes whilst the sample leg is being purged.

The sample leg is flushed to waste, and the needle valve which is situated well downstream of the sampling septum is slowly closed so that pressure in the sample line is brought up to main oil line or vessel pressure, almost all the pressure drop for the last 10 to 15 seconds of flushing occurring at the needle valve. In this way, the oil sampled through the septum fitting should be representative.

With the syringe valve open, and holding the plunger fully in, the needle is inserted through the septum into the centre of the sample line. The pressure in the line is allowed to gently push the plunger back, and the syringe is flushed several times. The twist valve is then closed, trapping about 50 µl of oil in the syringe. The needle is withdrawn from the septum, and the syringe plunger pulled back to the 90 µl mark, so reducing the internal sample pressure.

The choice of a 50 µl aliquot has to be a compromise between the need to precisely weigh the sample into the GC, and modern chromatographic practice of ever decreasing sample size.

Prior to analysis, the sampling needle is removed, the syringe washed with Freon, blown dry, and a clean needle fitted. If any leaks are noted, traces of oil behind the plunger tip for example, the sample is discarded and the syringe set aside for maintenance.

Weighing

The syringe fitted with needle and holding the sample is weighed on a balance capable of measuring 60g to 0.01 mg immediately before injection into the GC. As soon as the injection has been made and the needle has cooled to room temperature, the syringe is reweighed, the difference in weights being the weight of the injection.

Gas Chromatography

The column is 2.5 m of 1/4" o.d. thin walled stainless steel guarded by a 15 cm long 1/4" o.d. precolumn. Both are packed with solvent washed Porapak Q, the top third of the precolumn being left unpacked to provide a vapourisation chamber. This size of column allows for a sufficiently heavy column loading that the sample can be weighed into the machine with acceptable precision. See figures 3.

The helium carrier flow is set to 60 ml/minute. The injection port is held isothermally at 230 'C to ensure that despite the large injection size, all the components of interest are quickly vapourised.

A heated rotary valve is positioned between the precolumn and the main column to allow the precolumn to be backflushed to vent, this minimises the amount of C6 plus passing onto the main column. To obviate the risk of any pressure surge from the valve affecting peak detection, it is not fired until the gap between the propane and the isobutane peaks.

Separation efficiency should be monitored, and the precolumn withdrawn and repacked when the accumulation of heavy residue starts to cause peak degradation. When in continuous use, a precolumn lasts for 50 to 75 injections, and if it is left backflushing between samples about 100 to 200 injections. The freshly packed precolumn should be reconditioned in the backflush mode for several hours prior to use.

During the injection, the column oven is maintained at -50'C. This allows the injection to be made over a period of a few seconds, and the syringe to be flushed several times with carrier, so ensuring that all the components of interest are injected, without causing multiple peaks or an unacceptable decrease in plate number.

At -50'C, good oxygen / nitrogen separation can be achieved, and the oxygen figure is used to correct the apparent nitrogen content for air present in the syringe needle.

Once the oxygen has eluted (3 minutes), the column oven is temperature programmed at 25'C per minute to 130'C, and then at 5'C per minute to 250'C. Although it is not normally accepted as desirable to take Porapak Q above about 230'C, a final temperature of 250 'C gives a total run time of 35 minutes with no apparent baseline problems.

A thermal conductivity detector and a flame ionisation detector are used in series, the TCD operating at 0.5 mV full scale with a filament temperature of 340'C, drawing a current of about 210 mA. The FID is used at a sensitivity of 10^{-8} amp / mV. Both detector blocks are heated to 275'C. A carrier to fuel ratio of 1:1 and a fuel to air ratio of 1:5 appears to give optimum operation of the FID.

The nitrogen, oxygen, carbon dioxide and water are detected on the TCD, and all the hydrocarbons on the FID.

Although it was not initially anticipated that the method would cover the determination of water, a reasonably shaped and integratable water peak is observed, and this can be used to find the water content of the syringe so that the results can be expressed on a dry basis.

Determination of Light Ends in Unstabilised Crude Oils

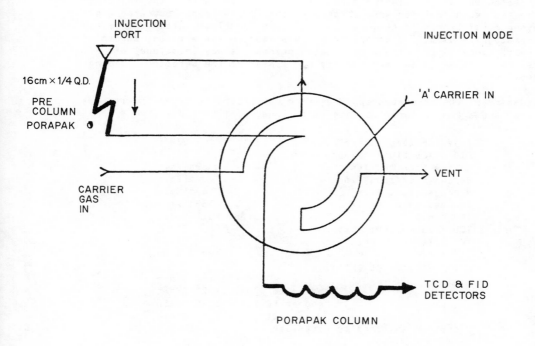

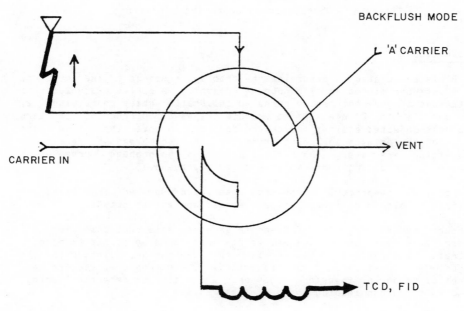

FIGURE 3

To allow for a sufficient accuracy in the weighing step, the injection size is such that for some components the FID is being operated well beyond its linear range, and this gave problems with the first chromatograph used. Once the change was made to the existing machine - a Varian model 4600 - the problem disappeared, since this model of FID has been found to be reproducibly non linear at the detector loadings used.

A summary of the chromatograph operating conditions is given in figure 4 below, and a sample chromatograph is included as figure 5.

```
Injector temperature                        230'C
FID & TCD block temperatures                275'C
Helium carrier flow                         60ml/min
Helium backflush flow                       30ml/min
Hydrogen fuel flow                          60ml/min
Combustion air flow                         300ml/min
TCD range                                   0.5mV full scale
TCD filament temperature                    340'C
TCD filament current                        210mA
FID range                                   10^-8 A/mV
Oven temperature profile
                                 -50'C for 3 min
                    program to 130'C at 25'C/min
                     program to 250'C at 5'C/min
                          hold at 250'C for 10 min
Precolumn switched to backflush             15 min
```

Figure 4   Chromatograph operating conditions.

Calibration

The GC is calibrated to determine the absolute response of the detectors in counts per microgram of component present. A series of volumetric injections of gravimetrically prepared calibration gases containing N_2, CO_2 and C1 to C4 are used to produce graphs of micrograms of each component injected against integrator counts. To cover the entire range of interest for each component several calibration blends of different composition are used, the highest points on the propane curve being obtained by injecting known volumes of pure propane.

The pentane response is found from a series of weighed injections of gravimetric blends of isopentane and normal pentane in octane.

Although in theory, The FID should show the same response for all isomers of a given carbon number, it has been found desirable to produce seperate calibration curves for isobutane and butane, since with the large injection sizes used in this method, the response factor for any weight of component injected is not only a function of detector loading, but also of column loading and hence peak shape.

The response for water is found from weighed injections of blends of water in dry methanol; although these are made up gravimetrically, they are checked by Karl Fischer.

A series of curves is produced of micrograms injected against integrator counts for each component, although with a microcomputer available it has been found better to express these as a series of polynomials.

Determination of Light Ends in Unstabilised Crude Oils

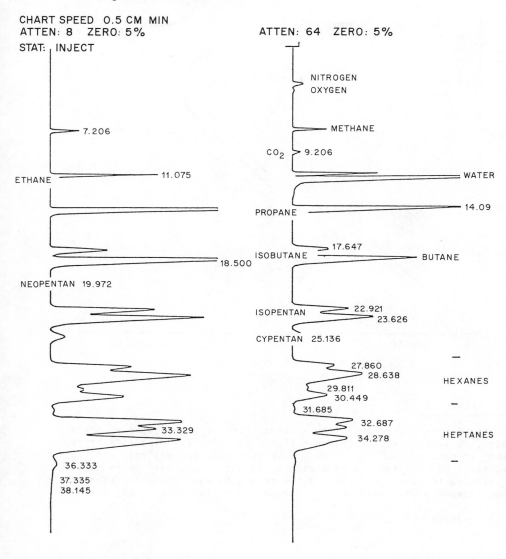

FIGURE 5

The reproducibility of the FID's non linear response is sensitive to the carrier : fuel : air ratios, and it is therefore desirable to use a machine with high quality pneumatic control systems, situated in a thermostatically controlled box.

The calibration normally remains valid until the machine is disturbed for servicing, or the gas regulator settings are altered, but a quick check on the calibration can be made by injecting a known volume of a gas standard mixture, and this should be done on a weekly basis or whenever the precolumn has been repacked.

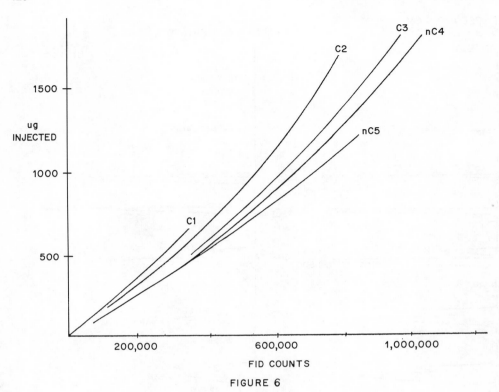

FIGURE 6

Syringe Integrity

A sample of unstabilised crude was taken into a 1 litre single phase cylinder, 8 syringe subsamples were then drawn from it, and were hence assumed to contain identical oil. Five syringes were analysed within the first 24 hours, and the remaining three within the next three days. From the results given in Table 1 below it can be seen that the samples did not appear to have deteriorated during the period of storage.

Table 1   syringe integrity

	First 24 hours average wt%	σ	Between 1 and 3 days average wt%	σ
C1	0.09	.008	0.10	.012
C2	0.31	.034	0.32	.021
C3	1.44	.061	1.47	.035
iC4	0.32	.004	0.31	.006
nC4	1.48	.029	1.46	.021
iC5	0.68	.018	0.67	.015
nC5	1.17	.038	1.13	.025
cyC5	0.12	.000	0.12	.006
total C1-C4	3.64		3.66	
total C1-C5	5.61		5.58	

σ = the standard deviation

Determination of Light Ends in Unstabilised Crude Oils

Comparison with the Flash Stabilisation Technique

One litre samples of unstabilised crude oil were taken into Welker single phase cylinders, and then subjected to both Possum and flash stabilisation analyses. In the latter, the oil was separated at controlled temperature and measured pressure conditions into gas and stabilised liquid fractions, these were then analysed by IP 345 and 344 respectively, and the results mathematically recombined to predict the original composition.

Table 2a   Flash stabilisation

Run	1	2	3	4	5	Mean	σ	cv
N2	0.02	0.00	0.01	0.00	0.01	0.01		
CO2	0.12	0.13	0.12	0.12	0.14	0.13	0.010	7.5
C1	0.36	0.38	0.38	0.34	0.42	0.38	0.030	7.9
C2	0.89	0.93	0.91	0.86	1.00	0.92	0.053	5.7
C3	2.90	3.03	2.96	2.82	3.19	2.98	0.141	4.7
iC4	0.60	0.62	0.61	0.63	0.64	0.62	0.016	2.6
nC4	2.12	2.20	2.16	2.25	2.24	2.20	0.054	2.5
iC5	0.91	0.94	0.92	0.99	0.94	0.94	0.031	3.3
nC5	1.41	1.46	1.43	1.51	1.45	1.45	0.038	2.6
cyC5	0.18	0.19	0.18	0.19	0.18	0.18	0.005	2.8

σ = standard deviation   cv = coefficient of variation

Table 2b   Possum

Run	1	2	3	4	5	6	7	Mean	σ	cv
N2	0.01	0.02	0.04	0.02	0.03	0.03	0.02			
CO2	0.14	0.14	0.13	0.13	0.13	0.13	0.13			
C1	0.39	0.38	0.38	0.38	0.37	0.39	0.39	0.38	0.008	1.8
C2	0.96	0.96	0.90	0.92	0.92	0.94	0.94	0.93	0.022	2.4
C3	3.13	3.05	2.88	2.93	2.93	2.97	2.98	2.98	0.087	2.9
iC4	0.61	0.57	0.57	0.57	0.56	0.58	0.61	0.58	0.020	4.9
nC4	2.52	2.45	2.35	2.38	2.34	2.43	2.42	2.41	0.062	2.6
iC5	0.90	0.88	0.85	0.84	0.84	0.89	0.89	0.87	0.026	3.0
nC5	1.46	1.43	1.39	1.39	1.37	1.43	1.44	1.42	0.032	2.2
cyC5	0.14	0.13	0.13	0.13	0.13	0.13	0.13	0.13	0.004	3.1

σ = standard deviation   cv = coefficient of variation

Table 2c   Summary

	Flash separation			Possum		
	mean wt%	standard deviation	coefficient of variance	mean wt%	standard deviation	coefficient of variance
C1	0.38	0.030	7.9	0.38	0.008	1.8
C2	0.92	0.053	5.7	0.93	0.022	2.4
C3	2.98	0.141	4.7	2.98	0.087	2.9
iC4	0.62	0.016	2.6	0.58	0.020	4.9
nC4	2.20	0.054	2.5	2.41	0.062	2.6
iC5	0.94	0.031	3.3	0.87	0.032	3.0
nC5	1.45	0.038	2.8	1.42	0.004	3.1

In another trial, thirteen single phase cylinder samples from each of several offshore installations were Possum sampled and analysed prior to being run through the normal flash stabilisation routine, and a comparison of the results is included in Table 3.

Table 3 Comparison of Possum and flash separation

	High GOR Platform		Low GOR Platform	
	Means w/w %		Means w/w %	
	Flash	Possum	Flash	Possum
N_2	0.00	0.00	0.00	0.00
CO_2	0.11	0.11	0.01	0.04
C_1	0.51	0.49	0.05	0.07
C_2	1.02	0.99	0.13	0.15
C_3	3.12	3.12	0.65	0.72
iC_4	0.61	0.68	0.18	0.21
nC_4	2.85	2.99	0.87	0.91
iC_5	1.13	1.14	0.51	0.50
nC_5	1.90	2.05	0.85	0.81
cy_5	0.29	0.27	0.12	0.11
$\sum N_2-C_4$	8.25	8.42	1.92	2.13
$\sum N_2-C_5$	11.57	11.89	3.42	3.56

n = 13

From the above it can be seen that there is better agreement between the Possum and flash methods for the platform with the higher Gas to Oil Ratio (GOR) than the lower one. This is thought to be largly due to the problems of measuring the small volumes of flash gas produced from the low GOR crude in the flash method.

Predicted vs Actual Product Make

Possum has been used as an operational hyrocarbon accounting tool during periods on a crude oil processing facility when normal sampling procedures were not available. The live and stabilised crude streams were analysed and the difference in compositions used to calculate the entitlement of participating companies to LPG products. The calculated LPG production makes using Possum data were within 5% of the metered quantities.

Discussion
==========

We feel that over the last seven years we have demonstrated that Possum is a fairly rapid analysis method with much simpler sampling requirements than either the onshore or offshore stabilisation methods. Minimal sample preparation is required in the laboratory other than the cleaning and weighing of the syringe. There is no possibility of sub sampling errors being introduced, and it has been found that once the chromatograph is calibrated, analyses can be carried out on a routine basis. The results given in the tables above were produced by about thirty different laboratory personnel over six years.

Unless the sample has been drawn from a thoroughly mixed pipeline, the water content found by the Possum method is unlikely to be representative of anything other than water in the syringe, but this water figure can be used to express the oil composition on a dry basis.

Whilst in theory, an internal standard method such as the Exxon method which has been modified by Shell, to allow for introduction of the sample using a liquid sample valve, should give greater analysis precision, it requires pressurised crude to be subsampled into a small cylinder with all the inherent problems of mixing crude samples under pressure, and the internal standard to be blended gravimetrically before again mixing and further subsampling. We would suggest that overall the Possum method, due to its being much simpler, probably has the better precision when the complete system from plant to result is considered.

Acknowledgements

I would like to thank my colleagues within Moore, Barrett and Redwood for their help in developing this method.

References

1. Abbott DJ and Wright CM in Petroanalysis 81 (ed.GB Crump) Wiley.

2. Rushby B in Petroleum Review Feb. 1986.

Petroanalysis '87
Edited by G. B. Crump
© 1988 John Wiley & Sons Ltd

Analytical Developments in the Monitoring of Atmospheres Associated with Oil Based Drilling Fluids

J.W. Hamlin*, M.H. Henderson and K.J. Saunders
* BP Chemicals Limited, Belgrave House, London SW1W 0SU
B.P. Research Centre, Chertsey Road, Sunbury-on-Thames, Middlesex TW16 7LN

INTRODUCTION

In the early 1980's a requirement was identified to conduct personal and fixed position time weighted average (TWA) monitoring of airborne hydrocarbons generated during the use of oil based drilling fluids in mud handling areas. Traditionally diesel oil was used as the base oil. However, with the increased use of oil based drilling fluids several companies focused attention on the development of environmentally acceptable "low toxicity" base oils, including BP Chemicals which developed BP 8313, a kerosine type product containing C_8–C_{16} hydrocarbons.

The methods available initially for the estimation of the atmospheric concentration of total hydrocarbons in the mud handling area included an instantaneous measurement and a method for determining the TWA concentration based upon sorption/desorption.

The instantaneous measurement was obtained using the Century Organic Vapour Analyser (OVA) Model 128. This is a portable and intrinsically safe direct readout instrument. Using flame ionisation detection, it gives either a continuous display of the total hydrocarbon levels or is used to identify trace level sources.

The TWA method originally available involved drawing the atmosphere at a known sampling rate, through a sorbent (Chromosorb 106) contained in a metal tube. Subsequently, the sorbed material was thermally desorbed into a packed column and analysed by gas chromatography (GC). The atmospheric concentration, as represented by the total area of the resulting chromatogram, was calculated by reference to standards. The limitations of this technique arose from incomplete desorption from the Chromosorb 106, the variability of response factors with carbon number and the poor resolution obtained from packed columns.

A technique which overcame the problems of variability of response factors and poor resolution, used capillary column GC technology, but was only suitable for hydrocarbons up to C_{10} (1). This method had been validated for TWA measurements of both fixed point and personal exposure to gasoline atmospheres.

This paper describes the extension of this latter method in which the sorbent in the sampling tubes and in the analytical apparatus is replaced with Tenax GC. The substitution of Chromosorb 106 with Tenax GC, with its reduced breakthrough volume facilitates recovery of up to C_{16} hydrocarbons (2).

EXPERIMENTAL

Analytical Apparatus

Thermal desorption was carried out using a Perkin-Elmer Automatic Thermal Desorber (ATD50). The standard injection assembly was replaced by a modified Varian Aerograph capillary splitter as shown in Figure 1. The analytical capillary column passed from the GC by way of the heated transfer line into the outlet of the secondary trap of the ATD 50 (Figure 2), which was packed with 40 mg of Tenax GC porous polymer. The GC and ATD50 operating conditions are given in Table 1.

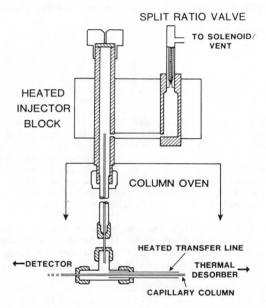

Figure 1.   Modified Varian Aerograph Capillary Splitter.

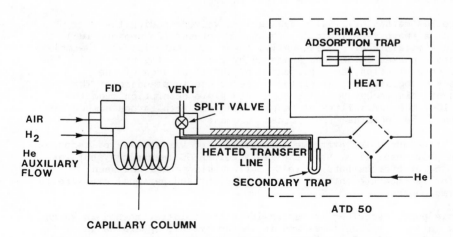

Figure 2.   Gas Chromatograph and Automatic Thermal Desorber System for Capillary Operation.

TABLE 1

GAS CHROMATOGRAPH AND AUTOMATIC THERMAL DESORBER OPERATING CONDITIONS

a) Thermal Desorber

Mode 2; two stage thermal desorption
Secondary trap contains 40mg Tenax GC

Primary desorption temperature	250°C
Primary desorption time	10 minutes
Secondary trap initial temperature	-30°C
Secondary trap final temperature	300°C
Transfer line temperature	150°C

b) Gas Chromatograph

Column	50 m x 0.32 mm ID. Fused silica. OV-1. Film Thickness 0.25 bonded
Carrier gas	Helium
Flow rate	1.5 ml/min at 10.5 psig
Make up gas	Nitrogen
Make up gas flow rate	30 ml/min
Split ratio	100:1
Detector temperature	300°C
Oven temperature	40°C - 10°C/min - 280°C
Amplifier range	10^{-11} amps

Sampling Tubes

Perkin-Elmer sampling tubes (90 mm x 6.4 mm stainless steel) were packed with 200 mg of Tenax GC 60/80 mesh conditioned for 16 hours at 300°C under a stream of nitrogen. The cleanliness of the tubes was confirmed prior to sampling using the normal ATD50 desorption cycle (Table 1). Tubes were accepted if the total peak area, expressed as hydrocarbon, was less than an atmospheric equivalent of 0.1 mg/m^3 and if the peak area of individual hydrocarbons was less than an atmospheric equivalent of 0.01 mg/m^3, for a 2500 ml sample.

VALIDATION

Laboratory

An indication of the precision, accuracy and the range of complete recovery was required from the laboratory validation of the developed method. This was achieved by preparing quantitative standards of the requisite materials, loading the sampling tubes and analysing them under the established conditions.

a) Preparation of Standard Blends

A 2.5% standard blend of $n\text{-}C_8\text{-}n\text{-}C_{16}$ hydrocarbons and selected aromatic hydrocarbons was prepared by accurately weighing approximately 0.25 g of each component into a 10 ml volumetric flask and then making up with spectroscopic grade cyclohexane. Solutions of 0.25%, 0.05% and 0.025% wt/vol were prepared by dilution of the master solution.

b) Loading and Desorption of Sampling Tubes

An aliquot (1 µl) of one of the standard blends was loaded onto the front grooved end of the sampling tubes using a 10 µl Hamilton syringe. The tube was then purged for 1 minute with a stream of gas flowing at 10 ml/min to deposit the blend onto the sorbent, before being capped with analytical end caps and placed in the carousel of the ATD50.

The ATD50 cycle was started and the contents of the tube desorbed into the gas chromatograph (Figure 3). The oven temperature programme and data collection were started automatically.

When the desorption and analysis cycle was complete, the tube was subjected to a second cycle under the same conditions to check for complete recovery.

This procedure was repeated for each of the blends.

Field

Subsequent to laboratory validation, the method was field validated under normal operating conditions in the mud handling area of two North Sea drilling rigs. As discussed above, the laboratory comparisons gave an indication of the recovery, accuracy and precision of the developed method. However, in the field, as no validated comparison method exists, it was only possible to obtain precision values from the data obtained.

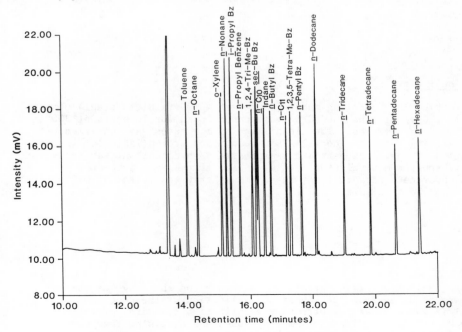

Figure 3.   Chromatogram of Standard Blend.

To obtain an estimate of this precision, side by side replicate sampling of the atmosphere was performed. Six sampling tubes were taped together ensuring that their ends were level with each other and each tube was connected to its own battery operated Accuhaler pump by a silicone rubber tube. Samples of the hydrocarbon containing atmosphere were collected over 12 hour periods at a nominal flow rate of 5 ml per minute. In all, 15 different locations within the mud handling area were monitored. The locations were carefully selected to give a broad spread of levels of airborne hydrocarbons and were not representative of the working environment.

In all cases, at the end of each sampling period the sample volume was noted and the tubes removed and capped. The tubes were subsequently desorbed on the ATD50 using the established conditions.

RESULTS AND DISCUSSION

Laboratory

Repeated injection of the standard blends showed good repeatability on desorption and essentially complete recovery up to and including $n-C_{16}$. A typical set of results for 14 separate injections is shown in Table 2. The responses, relative to n-undecane, all fall between 1 ± 0.12 and the coefficients of variation were in the range 10-15%. This level of precision was normal for the type of analysis and application. The second desorption of each tube produced very little extra hydrocarbon and although small amounts of higher hydrocarbons were found at the higher concentrations, these did not exceed 0.5% of the total quantity injected.

Field

The data obtained from the 9 fixed point locations on rig A are given in Table 3 and from 6 fixed point locations on rig B in Table 4.

The concentrations of the individual compounds present, together with the total hydrocarbon content, were calculated using the averaged laboratory determined response factor. By means of normal statistical techniques, the outliers were removed and the remaining results in each area averaged. Coefficients of variation of between 10 and 60% were seen over the range of the measurements. The mean coefficient of variation of the results from rig A was 29.3% and from rig B was 17.8% compared with a mean coefficient of 12.5% for the laboratory standards. This level is acceptable considering the type of atmosphere sampled. The total hydrocarbon results obtained ranged from 0.5 - 2770 mg/m^3 and the method performed well over the whole of this range.

The chromatograms of the atmosphere (Figure 4) displayed a typical kerosine type profile, very similar to the profile of the base oil, but with some discrimination towards the light hydrocarbons. This would indicate that the atmosphere contained not only vapour, but also mists and aerosols. If vapour only had been present, the chromatogram would have contained little hydrocarbon above $n-C_{10}$.

TABLE 2

STATISTICAL ANALYSIS OF 14 REPETITIVE INJECTIONS OF THE 0.25% HYDROCARBON BLEND

Compounds	Retention Time min	Weight in Standard μg	Recovery Mean Area	Coefficient of Variation %	Response Factor	Relative Response Factor
Toluene	8.60	1.4	10867	11.8	7762	1.07
n-Octane	9.27	1.4	9829	11.9	6909	0.95
o-Xylene	10.70	1.5	12193	12.6	8129	1.12
n-Nonane	10.95	1.8	13777	10.8	7654	1.05
i-Propylbenzene	11.29	1.9	15265	11.9	8034	1.11
n-Propylbenzene	11.83	1.4	10667	12.6	7619	1.05
1,2,4-Trimethylbenzene	12.58	1.5	11609	12.5	7739	1.07
n-Decane	12.81	1.7	13142	12.5	7731	1.07
sec-Butylbenzene	12.92	1.5	11294	12.3	7529	1.04
Indane	13.35	1.7	11914	12.7	7008	0.97
n-Butylbenzene	13.73	1.4	11402	12.6	8144	1.12
n-Undecane	14.63	1.3	9435	13.8	7258	1.00
1,2,3,5-Tetramethylbenzene	14.92	1.6	11797	12.6	7373	1.02
n-Pentylbenzene	15.55	1.6	10891	14.2	6807	0.94
n-Dodecane	16.38	1.9	15009	13.4	7899	1.09
n-Tridecane	18.05	1.4	10204	12.4	7289	1.00
n-Tetradecane	19.63	1.4	9620	12.1	6871	0.95
n-Pentadecane	21.14	1.2	8728	12.5	7273	1.00
n-Hexadecane	22.57	1.4	9398	12.6	6784	0.93

TABLE 3

MEAN CONCENTRATIONS OF MAJOR HYDROCARBON COMPONENTS AT VARIOUS LOCATIONS WITHIN THE MUD HANDLING AREA ON DRILLING RIG A (mg/m^3)

Location	1	2	3	4	5	6	7	8	9
Toluene	0.1	0.1	2.0			0.6	0.4	0.5	0.8
n-Octane		0.1	5.2			1.8	0.7	0.9	1.5
o-Xylene		0.4	6.5			2.2	0.3	0.3	0.5
n-Nonane	0.1	0.2	17.9		0.1	5.4	1.0	1.1	2.1
n-Propylbenzene	0.1	0.4	8.2			2.8	0.4	0.4	0.9
1,2,4-Trimethylbenzene	0.1	0.4	18.4			5.8	1.0	0.9	2.0
n-Decane	0.3	1.4	53.6	0.1	0.1	15.5	2.4	2.4	4.8
Indane	0.1	0.5	23.2		0.1	6.4	1.8	1.6	3.2
n-Butylbenzene							0.8	0.7	2.2
n-Undecane	0.6	3.3	133.3	0.1	0.1	35.1	5.0	5.5	10.9
1,2,3,5-Tetramethylbenzene	0.1	1.0	13.1			3.5	1.9	1.9	3.9
Naphthalene			54.8			11.5			
n-Dodecane	0.6	2.5	124.8	0.1	0.2	21.0	2.9	4.0	7.0
n-Tridecane	0.3	1.0	82.4	0.1	0.3	10.5	1.5	2.6	4.0
n-Tetradecane	0.2	0.4	42.3	0.1	0.3	3.8	0.4	1.0	1.3
n-Pentadecane	0.1	0.1	10.8	0.1	0.3	0.8	0.1	0.2	0.2
n-Hexadecane			1.5		0.1	0.1	0.1	0.1	0.1
Total hydrocarbons	5.0	26.5	1343	0.5	2.2	305	49.6	58.2	107.1
Number of Results	5	5	5	5	4	5	4	6	5
Coefficient of variation %	40.1	36.1	23.8	24.3	28.4	46.8	22.8	16.2	25.4

TABLE 4

MEAN CONCENTRATIONS OF MAJOR HYDROCARBON COMPONENTS AT VARIOUS LOCATIONS WITHIN THE MUD HANDLING AREA ON DRILLING RIG B (mg/m^3)

Location	1	2	3	4	5	6
Toluene	0.8	1.2	1.9	1.9	1.0	2.9
n-Octane	0.9	6.6	9.8	11.2	4.0	3.5
o-Xylene	3.8	5.1	8.1	12.6	5.0	4.1
n-Nonane	10.8	23.9	32.8	52.2	14.7	25.0
n-Propylbenzene	8.6	7.7	10.9	20.2	3.1	23.5
1,2,4-Trimethylbenzene	20.5	25.6	35.4	67.1	20.4	72.9
n-Decane	84.9	48.6	65.7	114.6	35.2	116.9
Indane	24.1	14.8	22.3	39.7	15.2	33.7
n-Butylbenzene	31.8	18.6	27.6	36.8	17.0	33.7
n-Undecane	324.9	90.7	122.7	216.1	72.6	181.6
1,2,3,5-Tetramethylbenzene	10.3	19.1	34.8	15.1	18.6	74.6
Naphthalene	69.8	18.9	25.7	30.0	16.4	83.9
n-Dodecane	227.9	60.2	80.1	145.8	47.7	121.1
n-Tridecane	151.1	25.2	45.2	72.5	20.3	108.9
n-Tetradecane	64.6	13.1	19.1	29.6	10.3	42.7
n-Pentadecane	16.7	2.7	4.0	6.1	2.0	22.3
n-Hexadecane	1.3	0.1	0.2	0.3	<0.1	2.4
Total Hydrocarbons	2394	797	1126	2067	679	2770
Number of Results	3	3	3	3	3	3
Coefficient of Variation %	10.2	24.2	19.6	17.6	17.3	18.8

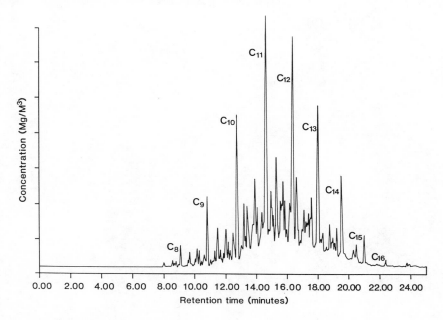

Figure 4.   Typical Chromatogram of Mud System Atmosphere.

CONCLUSIONS

Using the technique described above, it has been demonstrated that quantitative recovery of up to C_{16} hydrocarbons was obtained and that individual hydrocarbons could be resolved and quantified. The technique is suitable for measuring the airborne concentration of C_8-C_{16} hydrocarbons during the handling of oil based drilling fluids. The authors hope that the method will find general acceptance for both fixed point and personal monitoring during these operations.

REFERENCES

(1) Price, J.A. and Saunders, K.J., Determination of Airborne Methyl Tert-Butyl Ether in Gasoline Atmospheres, Analyst, 109, 1984, 829-834.

(2) Henderson, M.H. and Saunders, K.J., The Determination of C_7-C_{16} Hydrocarbon Vapours in Kerosine Atmospheres, Am Ind Hyg Assoc J, 48, 1987, in press.

Automated Identification of Petroleum Refinery Streams

D.J. Abbott
Esso Research Centre, Milton Hill, Abingdon, Oxon OX13 6AE

SUMMARY

A method for identifying refinery streams automatically by pattern recognition is described. The method is based on capillary gas chromatography giving peak area data and simulated distillation data simultaneously.

INTRODUCTION

In a large petrochemical complex there are usually many different crude oils, products and intermediate streams. In order to locate the source of leakages, any hydrocarbons found in cooling water or sewer systems are usually analysed by GC to obtain a 'fingerprint', which is then compared with a library of 'fingerprints'. If the sample is a pure stream, a skilled operator can usually narrow down the possibilities to just a few from the library. In view of the continuous operation of refineries, automatic identification of unknown samples by a computer system would be advantageous.

There are several methods available for pattern recognition by computer. Probably the most familiar to analytical chemists are the programs used in mass spectrometry for matching a mass spectrum against the various libraries of mass spectra. There are also several different methods of pattern recognition which have been applied to weathered crude oils(1). However, all these methods rely on resolved peaks being obtained. Many refinery samples, especially lube oils and fuel oils, will not give resolved peaks even with capillary columns, and these methods are therefore not applicable to pattern recognition in a refinery.

When a human operator performs pattern recognition he looks at essentially two sets of data; 1) the relative ratios of easily identified components, and 2) the overall distribution of the components of the samples. It is this second set of data which allows samples such as lube oils and fuel oils to be recognised. The overall distribution of sample components is of course measured by distillation, and simulating distillation by GC has been carried out for many years(2). Many computer data systems allow peak areas and simulated distillation data to be determined simultaneously, and these are ideal for this application.

EXPERIMENTAL

Analyses were performed on a 25m x 0.25mm i.d. capillary column coated with methylsilicone (CP5, Chrompack). The column was programmed from $40°$ to $290°$

TABLE 1

Pattern Recognition Data for Eleven Samples

Sample	Cumulative Percent Off							Peak File Data		
	5 °C	10 °C	30 °C	50 °C	70 °C	90 °C	95 °C	n-paraffin mass % C10 to C20	ratio nC15/nC10	ratio nC15/nC20
1	69	73	93	137	179	230	240	1	0	0
2	146	160	182	201	223	247	253	31	0.2	0
3	76	83	101	119	138	169	197	0	0	0
4	182	204	252	285	316	355	369	53	4.5	1.6
5	194	207	235	265	298	344	360	21	999	1.4
6	385	392	409	421	434	451	458	0	0	0
7	146	147	155	163	177	212	232	11	0	0
8	71	77	105	183	273	379	419	12	0.6	1.3
9	71	76	135	232	314	399	430	12	0.8	2.4
10	322	341	380	406	428	457	467	5	999	0.1
11	163	182	225	269	317	373	394	28	1.8	1.2

at 5°/min. The carrier gas was helium at 2ml/min. Injection was by autosampler through an inlet splitter (100:1) at 350°C. Peak areas and time slice areas were acquired simultaneouly by a Perkin-Elmer Sigma 10 data system and passed to a Commodore 8032 computer for calculation of simulated distillation data.

RESULTS AND DISCUSSION

Initially, a selection of eleven samples, including crudes, intermediate streams and final products, were analysed. Seven simulated distillation data points and three points from the peak file data were selected for pattern recognition purposes. The peak file data points are based on the identification of n-paraffin in the C10 to C20 region. One point is the sum of the mass percents of each n-paraffin between nC10 and nC20, the other points are ratios derived from the mass percents of nC10, nC15, and nC20. In the results, Table 1, 999 is used to indicate infinity. These samples are mostly fairly easy to distinguish visually from their chromatograms. However, samples 8 and 9 (both crude oils) are similar and samples 4, 5 and 11 (middle distillates) have many similarities. A simple pattern recognition program running in the Commodore 8032 computer easily identified any of these samples when run as an unknown.

A second test using very similar light streams was undertaken and the results are given in Table 2. Again the computer pattern recognition successfully identified any sample run as an unknown. It was of interest to see the effect of widening the parameters for pattern recognition and the results are presented in Table 3. Sample E31 was run as an unknown and matched with library data using first, a spread of +/- 5° on the simulated distillation data, and then a spread of +/- 10°. The wider spread brings only one other sample into contention as a possibility, showing the usefullness of the technique.

TABLE 2

Pattern Recognition Data for Light Streams

Sample	5°	10°	30°	50°	70°	90°	95°	C10 to C20	C15/C10	C15/C20
	\multicolumn{7}{c}{Cumulative % off}	\multicolumn{3}{c}{Peak File Date}		
Gasoline	69	73	93	137	179	230	240	1	0	0
LCN	95	95	99	102	106	119	128	0	0	0
E31	97	99	112	120	142	168	173	5.5	0	0
P115	97	99	107	116	128	150	157	0	0	0
PS1	96	98	106	114	123	141	148	0	0	0
Naphtha	96	97	103	113	135	162	170	1.6	0	0

Table 3

Pattern Recognition Tests

A) Plus/Minus 5°C

Best Matches	% Fit
E31	100
P 115	50
Naphtha	40
PS 1	40

B) Plus/Minus 10°C

Best Matches	% Fit
E31	100
Naphtha	90
P 115	60
PS 1	60

Clearly, other data points such as the ratios C17/Pristane and C18/Phytane could be added, and indeed would be necessary to help to distinguish between crude oils. Overall, something like 25 data points from both simulated distillation and peak files could easily be obtained and should be sufficient for most refereries.

REFERENCES

1) K. Urdal et al, Marine Pollution Bulletin, 17, 366. (1986)
2) R.D. Butler in 'Chromatography in Petroleum Analysis', edited by K. Altgelt and T. Gouw, M. Dekker, New York. (1979)

Oil Analysis and Machine Condition Monitoring Techniques

R.F.W. Cutler
Wearcheck Laboratories, Robertson Research International, Llandudno, Gwynedd, LL30 1SA

1 INTRODUCTION

The idea of condition monitoring by spectrometric analysis of used oils, is not new! Indeed it was commercially available to industry in North America as early as the 1940's. The railways used it in a preventative maintenance programme to solve the problem of short engine life, but it was developed originally by engine builders and oil companies as a research and development tool.

Machine manufactures, oil companies and independent analytical laboratories introduced spectrometric oil analysis programmes (SOAP) to Britain in the 1960's and 70's.

The idea has now spread so successfully, that it is virtually impossible to sell mobile plant and lubricants without offering SOAP, as either a loss leader or as a condition of extended warranty.

The principle of SOAP is quite simple: engine and machine components are manufactured with different metals or alloys, eg aluminium pistons, chromium plated rings and copper/lead bearings.

Machine components wear, even in normal operating conditions. By monitoring the concentration of the various metal debris a set of baselines can be established. If regular condition monitoring is employed a sudden rise in any one or more of the metals would indicate a component is wearing abnormally.

Experienced diagnosticians can use this information to predict a problem before it becomes critical and long before it becomes evident to maintenance engineers.

It is logical therefore, that to protect a fleet of trucks or other machinery the service must:

1   Be reliable and offer a rapid turnaround time (24 hours).

2   Be used to monitor all the engines or machines in a fleet and not just a random selection. Such practice would amount to the majority of the units not being monitored. This would defeat the object of the excercise, but reduce the cost to the companies offering a free service and who advocate this idea.

3   Ideally it should be independent of both oil suppliers and machine manufacturers unless these supply companies themselves use on independent laboratory for this work.

    Many do infact accept this arguement and benefit tremendously by improved sales. A result of growing confidence in the companies prepared to allow their products to be assessed by independent analysts.

    It is simply an extension of the natural logic of people to believe in an independent unbiased assessment, to give a balanced view of the situation.

2   THE ECONOMICS OF CONDITION MONITORING

In the modern world of vastly expensive trucks and mobile plant, down-time and warranty claims are simply too expensive for all concerned.

Examples are shown below:
(i)   Mobile Plant equipment can cost well over   £200000
(ii)   Replacement engines and transmission boxes cost up to £25000
(iii)   Lost production in the mining industry costs up to   £50000/day
(iv)   To own one truck costs over   £200/day
(v)   Savings in truck maintenance costs can be   £200/vehicle/year
                                                               (50000 miles/year)

It is therefore essential for managers to be aware of abnormal conditions before they become critical problems.

Oil Analysis and Machine Condition Monitoring Techniques

Condition monitoring through oil analysis is an excellent way to achieve this in units such as engines, transmissions and other oil filled systems.

3 TECHNIQUES EMPLOYED IN CONDITION MONITORING

Condition monitoring is more than just monitoring wear metals. Comprehensive and reliable programmes are made up of three parts:
(a) Spectroscopic metal analysis
(b) Physical and chemical oil tests
(c) Interpretation and diagnosis of data

Specific oil tests and spectrometric metal analysis are necessary to evaluate the true machine and oil condition.

The ideal combination of tests are given below, with the information to be gained from each. It will be evident that the purposes of the tests tend to overlap other tests. This is deliberate action, to ensure results of one test is confirmed by data from another. The work is repeated if this is not the case.

Test required:

Wear and Additive Metal Analysis
Viscosity
Fuel Dilution
Oil Condition Index OCI
Dispersancy
Water/Glycol/Antifreeze
Environmental Dirt
Total Base Number
Particle Count

3 (i) SPECTROSCOPIC METAL ANALYSIS AND SIGNIFICANCE

Metals Determined	Significance
Barium }	Additive metals
Calcium }	Oil type
Magnesium }	Contamination by other oil
Zinc }	

Sodium }	Additive metals
Silicon }	Coolant contamination
Boron }	Dirt contamination
Aluminium	Upper cylinder and bearing wear
	Accessory drives
	Thrust washers
Chromium	Ring and seal wear
	Hydraulic rod wear
Molybdenum	Additive metal
	Ring and seal wear

Metals Determined	Significance
Copper	Bearing and bushing wear
	Thrust washers and clutch discs
	Cooler and turbo wear
Lead	Bearing and cooler wear
	Corrosion
	Petrol contamination
Tin	Bearing and liner wear
Manganese	Wear to steel components
Titanium	Environmental contamination in special cases
Nickel	Compressor tube wear
	Special steel component wear
	Fuel contamination
Silver	Bearing and liner wear in special cases
Vanadium	Valve stem wear
	Special steel component wear
	Fuel contamination

Oil Analysis and Machine Condition Monitoring Techniques

Typical Wear Metal Guidelines (examples only)

Table 1

Metal	N up to	C up to	S over
Iron	45	95	96+
Copper	15	45	46+
Aluminium	8	16	17+
Chromium	5	15	16+
Silicon	20	40	41+
Lead	25	70	71+
Tin	8	15	16+

The level of wear metals used to assess abnormal conditions differ for each engine or unit type, indeed they differ slightly for particular units of one type.

The information required must, therefore, be built up for each unit by regular monitoring, but it is possible to develop guidelines for a given unit type. This requires extensive field trials and large computer data files. Nevertheless, regular condition monitoring on a particular machine ultimately relies, not on the actual metallic values, but on sudden increases from the average.

Typical curves are shown below:

Figure 1

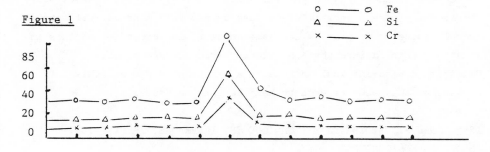

OIL MONITORING INTERVALS (MILES OR HOURS)

(ii) The analytical techniques used to monitor metal levels usually involve one of the following:

    (a)   ATOMIC ABSORBTION SPECTROPHOTOMETRY   (AA)
    (b)   INDUCTIVE COUPLED PLASMA EMISSION SPECTROSCOPY   (ICP)
    (c)   ROTRODE SPARK EMISSION SPECTROSCOPY   (RES)

The above techniques do not measure the total concentration of the metals present. Typical results obtained are shown in table 2. These are values determined on the same oil by the different techniques.

The variations are due to the interferences inherent in used oil analysis.

The prime reason for the differences in table 2 is particle size interference. This principle is adequately explained in the Analytical Chemistry Paper entitled "The Analytical Approach" Vol 56 No 9 August 1984.

In brief the arguement is that AA and ICP can only detect particles less than 2 microns and RES less than about 8 microns. Generally, used oils contain wear debris particles ranging in size from <1 micron to >30 microns. In some cases probably much larger. It therefore follows that the techniques normally employed, only detect a fraction of the total metals present.

This phenomenon, therefore, begs the question "how can the techniques reliably monitor abnormal wear, if the true concentration is not determined?"

The answer is that all the particle sizes increase with increased wear. Therefore, a rise in particles of less than 2 or 8 microns, reflects an overall increase in wear by the same amount. The exception to this is a catastrophic failure when only large particles are produced. Condition monitoring would not be in a position to prevent this situation, but could be used to help detect the cause.

TREND ANALYSIS BY THE SAME TECHNIQUE IS, THEREFORE, THE KEY TO CONDITION MONITORING.

Table 2

Metals	AA	ICP	RES	Wet Oxidation AA
Iron	91	90	147	210
Copper	44	96	125	195
Aluminium	3	3	7	9
Chromium	9	15	24	33
Molybdenum	4	6	14	31
Silicon	33	41	64	68

OTHER INTERFERENCES IN SPECTROMETRIC METAL ANALYSIS OF USED OILS

Inter-Element Effects

These are vitally important because a high concentration of certain metals enhance or depress the value of other metals. It is possible, therefore, to obtain a false value in these cases. However, spectroscopic interferences of this nature can be programmed out by computing the effects during the calibration and setting up procedures.

Matrix Effects

These are caused by differing chemical compositions of the used oils and standards available. They are more apparent with RES because no dilution is involved. Dilution effects are demonstrated by the rather extreme case in table 3.

Table 3

Metals	A	A1	B	B1	C	C1	D	D1
Copper	910	1080	615	270	420	137	405	140
Sodium	950	910	590	425	300	340	320	350

Values calculated to 100%

Two samples of Invert emulsion were each divided into four parts:
Parts A & A1 were analysed by RES, as received
Parts B & B1 were diluted to 25% with base oil before analysis by RES
Parts C & C1 were diluted to 5% with base oil before analysis by RES
Parts D & D1 were wet oxidized before analysis by AA

Standards for AA1, BB1 and CC1 were Conostan S-21 500
Standards for DD1 were aqueous primary standards

Note that in these cases the copper and sodium were in true solution, hence particle size played no role in this experiment.

It was evidence, that as the water concentration was diluted with mineral oil, the matrix became increasingly more similar to the matrix of the standard oil. This reduced the enhancing effect of the water and emulsifiers to give acceptable results.

To reduce matrix effects when employing RES, calibration standards are prepared from Conostan primary standards in both base oil and new lubricating oils.

Physical and Chemical Tests

The three major causes of machine failures are **fuel**, **water** and **dirt** contamination. It is essential the monitoring programme includes tests to detect all three major problems.

Test Description and Significance

Viscosity

The resistance to flow is measured by the time a given volume of oil flows under gravity between two points in a calibrated viscometer tube. Viscosity is essential to establish the correct grade of oil is in use, eg SAE 10, 20 or 30 etc. Typical changes in viscosity is shown in Figure 2. A = single grade, B = multi-grade, C & D = abnormal condition.

Slight changes are to be expected during use, but abnormal changes must be detected and reported. Abnormal changes are indicative of one or more of the following:

Overheating, fuel dilution, excessive use, polymer shear, high insolubles, wrong oil grade, general contamination

Figure 2

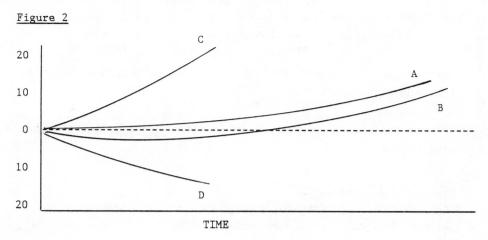

Viscosity change of 10% is considered abnormal. Depends on time oil has been in use.

Fuel Dilution

This is measured by either distillation or flash point. This test is essential to detect over rich mixtures, faulty injector systems and leaking pipework etc. An excess of fuel can lead to poor lubrication and excessive wear or failure. 5% is considered abnormal.

Oil Condition Index: OCI

The OCI is a measure of the conductivity of the oil. It indicates the concentration of soot and other conducting materials present in the sample, eg water and metal particles. An arbituary scale of 0-40 is used. A value of 10-12 is typical of new oils and values up to 28 are typical of used oils in good condition. Higher values indicate a problem may exist.

Again to establish the cause of an increase in the OCI value, other tests must be carried out.

Dispersancy

This is relevant to engine oils only, and is simply assessed by the blotter spot method.

Engine oils contain detergents and dispersants to disperse soot and other insoluble carbon residues throughout the oil. If insufficient is present the solids will coagulate to form sludges. These will tend to block passage ways etc and prevent lubrication of vital components.

Generally, an even distribution of soot is a good sign and a non-even distribution is a poor sign.

This information is used with the other data to assess the condition of the oil.

GLYCOL - ANTIFREEZE - WATER

Water can readily be detected by the crackle test. Indeed, qualitatively, the crackle test in an extremely reliable test. Simply by dropping oil onto a hot-plate at over 100°C will cause the water, if present, to crackle. No crackling noise, confirms no water is present.

Where water is detected it is measured by distillation or other means to establish its concentration, >0.2% is considered significant.

Having established water is present, it is necessary to identify its origin.
The options include:
1   Coolant contamination
2   Condensation
3   Supply tank contamination

If a coolant leak is the cause then glycol may be present. To check for glycol (ethylene glycol type antifreeze) the Periodate/Schiffs Reagent test is carried out, but this has many draw backs. If glycol is present the test will detect it, but interferences are possible from some oil additives. Thus a positive test is possible when none is present. However, the converse is also true! The usual case involves a small coolant leak where it is possible for the water to evaporate and the glycol to either do likewise or oxidize to the corresponding acid. Thus the tests will prove negative when a coolant leak actually exists.

Glycol present in engine or transmission oil is very serious. Limits of <50 ppm are given for particular sensitive units. The problem is overcome by measuring the metals used in the corrosion inhibitor package.

These packages contain combinations of sodium salts, silicates and borates. Not all may be used, but at least one will be present, and can be readily detected by spectroscopy. These metals are, therefore, monitored specifically during the general wear metal analysis programme.

ENVIRONMENTAL DIRT

The third major cause of engine or machine failure is dirt contamination. It acts as an abrasive on pistons rings, liners and bearings, etc to cause severe problems.

The type of dirt depends on the material being handled and the atmospheric working conditions. Generally, earth moving equipment and road vehicles are in contact with silicate dirt. Special processes, however, may involve titanium, limestone or coal as the major dirt component.

It is, therefore, essential to be aware of these factors during the diagnosis. The programme of metal analysis includes 20 elements which cover these possibilities.

Silicon limits are indicated in table 1, but the general wear metal levels are used to detect the effect of the contamination.

TOTAL BASE NUMBER

To protect the crankcase from acid attack by corrosive acids produced in the combustion chamber, the lubricant contains a degree of reserve alkalinity. This reserve alkalinity is expressed as mg of KoH/gm of oil and is described as the Total Base Number (TBN).

It is measured by one of two methods adopted by the Institute of Petroleum under the numbers IP177 and IP276.

The essential difference is that IP177 uses hydrochloric acid to neutralize the remaining alkali and IP276 uses perchloric acid in truly non-aqueous conditions. Both measure the end points by potentiometric means.

The data below shows the values obtained on three oils using the different techniques:

Table 4

	TBN	
	IP177	IP276
New Oil A	6.3	6.9
Used Oil A 100 hours	6.5	6.9
200 hours	5.1	7.1
300 hours	4.0	6.8
New Oil B	21.00	22.00
Used Oil B 100 hours	19.60	23.00
300 hours	17.30	24.20
700 hours	12.80	31.10
New Oil C	12.6	14.1
Used Oil C 72 hours	11.8	13.9
200 hours	10.0	14.8
350 hours	8.5	15.6

The data in table 4 shows the results determined by IP177 decreased with oil use. This is consistent with the reserve alkalinity being used up by mineral acids produced in the combustion chamber. However, table 4 shows that the results determined by IP 276 on the same oils, remained either constant or actually increased in value. This is totally inconsistent with the expected result. The arguement therefore remains as to which technique is valid. In general, oil companies quote IP276 values and this has encouraged engine builders to do likewise.

This causes a serious situation because some engine builders recommend oil changes when the TBN by IP276 falls to 50% of the original value. Table 4 confirms, in certain cases, the 50% decrease by IP276 will never occur, although the reserve alkalinity may have fallen significantly.

It was important to establish which one gives the realistic answer. It was argued that by controlling the process of neutralization of a new oil by sulphuric acid, it would be possible to determine which method gave the realistic value.

The Wearcheck laboratory carried out the following experiment:

The TBN values by both methods were determined on a sample of new engine oil. A further sample of the engine oil was treated with sufficient 0.1 N sulphuric acid to neutralize the equivalent of 6 mg/KOH/g.

The TBN values by both methods were determined on portions of this treated oil.

The results are shown in table 5

Table 5

	TBN (IP177)	TBN (IP276)
New Oil	9.7	11.3
Treated Oil	3.4	9.1
Expected Value	3.7	5.3

The experiment confirms that IP276 produces a new oil TBN value of approximately 2 mg KOH/g higher than the IP177 value. Table 5 shows that IP177 method produced the more realistic answer. Although in this case the TBN by 276 did fall.

The rise in TBN by IP276 in a real situation is accounted for by the following arguements:

1 The sulphuric acid produced is partly neutralized by engine metals as well as the oil additive base components. These additional metals were not present in the experiment above.

2 The metals of these salts and those of the neutralized alkali additives are "seen" by the much stronger perchloric acid, as basic compounds.

3 The reaction between perchloric acid and these neutralized salts will produce the corresponding mineral acid. However, in the non-aqueous phase, existing in the IP276 method, these mineral acids do not ionize readily to produce hydrogen ions. Therefore, behave similar to weak acids, which do not significantly effect the pH value. The perchloric acid nevertheless ionizes to a much greater extent and dominates the equilibrium, analogous to that of a strong and weak acid system.

The outcome is that, as the perchloric acid is neutralized by true basic compounds and these already neutral salts, (including those of engine metals) it produces a result to indicate the TBN is still high or increasing. However, the neutral salts would not be able to neutralize sulphuric acid, in a real situation. The IP276 method therefore gives a misleading value in used oil analysis.

The conclusion is that IP276 is an unacceptable method to employ in condition monitoring programmes and that IP177 should be used.

Particle Count

This test is generally applicable to hydraulic oils, but some companies are employing it to monitor transmission oils.

It is an excellent way of monitoring the cleanliness of an oil in terms of both metallic and non metallic debris. The importance for hydraulic oil is due to the fine tolerances of the oil ways and the risks of impairing the performance of the machines, eg needle valves and seal efficiency.

The techniques employed measure the total particles greater than 5 microns, 15, 25, 50 and 100 microns in 100 ml of sample.

The numbers obtained are large and cumbersome making it difficult to appreciate the cleanliness of the oil. A scale has therefore been devised by the International Standards Organisation to provide an index rating.

Oil Analysis and Machine Condition Monitoring Techniques

The tables below explain how this indexing is determined:

Table 6a

Particles	>5	>15	>25	>50	>100
No per 100 ml	110000	3500	1400	250	75
Total	115225	5225			
Range Number ISO SC Code	17	13			

Table 6b

Particles		Range
Greater	up to	
100000	2000000	21
250000	500000	19
64000	130000	17
16000	32000	15
4000	8000	13

The ISO SC Code range offers a simple expression of the total number of particles >5 microns and >15 microns. It reflects an over picture of the contamination distribution of particles >5 microns.

The choice of 5 and 15 microns also reflects the most damaging particle sizes. It is argued that hydraulic tolerance levels between moving components fall into two sizes, 5 microns and 15-40 microns. Particles larger than these sizes do not enter the system and those less than 5 microns pass through easily. However, those close to these dimensions, squeeze between the moving components and damage the surfaces.

Limits to the levels of dirt and particle size ranges depends on the sensitivity of the machinery and pressures involved. Figure 3 reflects the general trends expected for various oil systems.

Figure 3

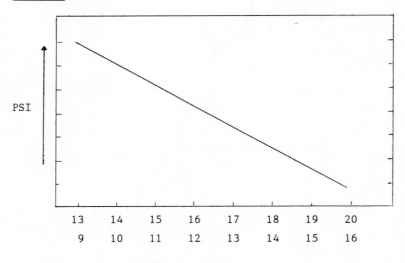

PARTICLE COUNT ISO CODE

There are many case studies available to admirably demonstrate the benefits of regular condition monitoring.

However, for this paper, two special ones have been described to show the advantages offered, to both machine manufacturers and end users.

4   Case 1

In this case a new lift system was developed for use in small hotels. Many lifts were installed within a few months of each other and were expected to operate for more than 35000 starts before an overall.

After a short time three failures occurred within a short space of time. An investigation proved the same problem existed in all cases involving excessive wear of the gear teeth. The reduction in metal allowed the gear teeth to shear off causing the lift cage to fall freely.

A reconstruction of the system was carried out under laboratory conditions. Regular oil analysis was carried out during the experiment to monitor wear rates. These were compared with the actual percentage loss of gear teeth metal.

The graphs below show how spectroscopic metal analysis could predict the extent of wear. Thus demonstrating how, if used, it could have prevented the accidents.

Figure 4

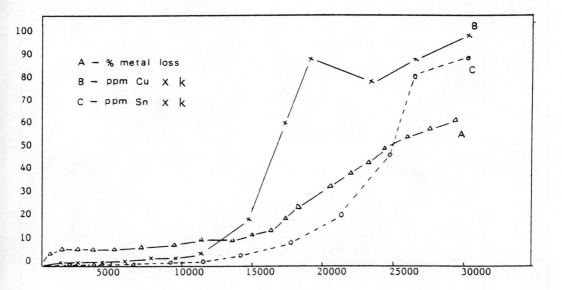

The problem was solved by increasing the size and altering the metallurgy of the gear system.

Case 2

Regular condition monitoring produced the following data:

LAB NUMBER	OIL CONDITION								WEAR METALS								
	Visc CSt at 40°C	FUEL	O.C.I	DIS PERS ANCY	WATER	GLYCOL	SODIUM	DIRT SILICON	ALUM INIUM	CHRO MIUM	COPPER	IRON	LEAD	TIN	Mo	Mg	B
194955	108 N		21 P	30 S			140 S	35 C	26 S	16 S	110 S	170 S	80 S	19 S	1 N	140 N	46 S

Sample No.
Sample Date 19/5/1986
Received Date
Unit Age 1282 hrs
Oil Age 120 hrs

SERIOUS COOLANT CONTAMINATION EVIDENT. WEAR NOW SERIOUS. ADVISE TAKE ACTION.

| 192614 | 106 N | | 20 P | 20 C | | 70 S | 29 C | 14 C | 11 C | 36 N | 85 C | 50 C | 11 C | 1 N | 150 N | 30 S |

Sample No
Sample Date 3/5/1986
Received Date
Unit Age 1162 hrs
Oil Age 230 hrs

COOLANT CLEARLY EVIDENT. WEAR INCREASING. ADVISE TAKE ACTION A.S.A.P.
_____(TELECON INFO)_____

| 189987 | 107 N | | 21 P | | N | 38 C | 21 C | 9 C | 7 C | 22 C | 40 N | 33 C | 9 C | 2 N | 170 N | 18 C |

Sample No
Sample Date 1/4/1987
Received Date
Unit Age 932 hrs
Oil Age 200 hrs

COOLANT PRESENT CAUSING WEAR TO UPPER CYLINDER AND BEARINGS. ADVISE TAKE ACTION AT NEXT SERVICE.

| 188463 | 106 N | | 19 G | N | N | 7 N | 10 N | 2 N | 1 N | 10 N | 70 N | 14 N | 2 N | 1 N | 150 N | 2 N |

Sample No
Sample Date 9/3/1987
Received Date
Unit Age 732 hrs
Oil Age 231 hrs

OIL CONDITION AND WEAR METAL LEVELS APPEAR SATISFACTORY. ADVISE NO ACTION.

| 187216 | 107 N | | 19 G | N | N | 8 N | 10 N | 3 N | 1 N | 8 N | 27 N | 10 N | 2 N | 1 N | 150 N | 2 N |

Sample No
Sample Date 3/3/1986
Received Date
Unit Age 420 hrs
Oil Age 150 hrs

OIL CONDITION AND WEAR METAL LEVELS APPEAR SATISFACTORY. ADVISE NO ACTION.

OIL BRAND/GRADE

UNIT HISTORY

The information shows there was a coolant leak into the engine sump after only a few working hour.

The owner was informed of this problem as a consequence of employing regular oil condition monitoring.

The results were reported to the manufacturer who promptly investigated the claim.

All his tests proved negative. There was no evidence of a leak after exhaustive testing and inspection. Nevertheless, the condition monitoring results continued to prove there was a leak.

After three consistent reports and deduction by elimination, the manufacturer concluded the problem was in the oil cooler, although nothing could be detected. Further rigorous tests on the cooler under laboratory conditions at operating temperatures showed the cooler was leaking. A single soldered joint was shown to be faulty, demonstrating that condition monitoring had successfully detected an otherwise undetectable fault.

This was proof of a justified warranty claim by the machine owner for maintenance repairs, which the manufacturer accepted.

However, further analysis by the Wearcheck laboratory was able to prove the metallurgy of the brazing material of the leaking joint differed from the other joints. This indicated the coolers were reconditioned components. The cooler supplier had alleged they were new coolers but the Wearcheck reports were sufficient proof for a successful claim by the plant manufacturer against the original supplier of the coolers. This protected the manufacturers' good name and the expense of a series of expensive warranty claims.

Wax Chromatography—The 80's Crossroads

A. Barker
Dussek Campbell Limited,

Modern quality assurance of waxes and wax blends requires a knowledge of the molecular weight distribution in the form of the Carbon Number Distribution of alkanes. (See Fig. 1) Unlike G.L.C. Simulated Distillation techniques, the resolution of the alkanes is very important to wax chromatography. During the G.L.C. separation of waxes the straight chain alkanes are eluted in order of molecular weight (or carbon number). In between the straight chain alkanes are much smaller peaks, which are mainly branched chain alkanes (with a small percentage of cycloalkanes and aromatics). For the past five years, the author has been developing quantititive G.L.C. as a means of characterising waxes and wax blends.

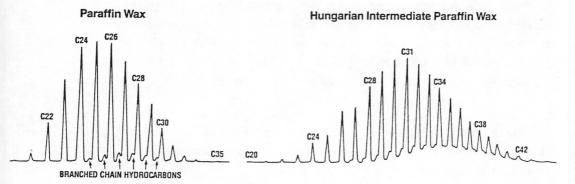

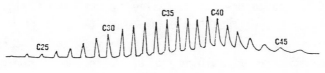

Fig. 1

Prior to 1980, dual packed column G.L.C., using flame ionisation detection (F.I.D.) was the standard method for evaluating the carbon number distribution of waxes. In Fig. 1 the three main types of petroleum wax can be seen separated by packed column G.L.C. into straight and branched chain alkanes, with the resolution decreasing as the carbon number increases. Several problems are associated with the packed column G.L.C. methods.

1. The lack of resolution of the alkanes means that the amounts of straight and branched chain molecules cannot be accurately measured (important in formulation).

2. It is impossible to directly formulate wax blends using the carbon number distribution analysis of the constituent raw materials.

3. Wax blend producers and customers use a great diversity of quantitative analytical methods. Analysts use different column sizes, stationary phases, temperature programmes, final column temperatures and sample preparation. Also there are problems in harmonisation of results due to differing methods of integration of peak areas and carbon number calculation.

In 1983, the author started background research in quantitative carbon number distribution analysis using a modified Perkin Elmer F33 G.L.C. and a 1.2m x 0.2mm, stainless steel column packed with 5% OV-1 or Dexil 300 on Chromosorb WaW. Modern diaphragm flow controllers were used to accurately control the O.F.N. grade nitrogen carrier gas.

Initially, there were problems of stationary phase degradation at $350^{\circ}C$ (with deterioration of peak resolution and column matching), variations in relative responses of the alkane series standard, and variations in results due to different modes of injection. Co-operative research with Mr. J. Roberts, of Phase Separation Ltd., led to stationary phase degradation at $350^{\circ}C$ being greatly curtailed by the use of high purity nitrogen (Air Products certified 5.5 grade), and by a careful lengthy conditioning procedure. This enables the author to constantly temperature programme packed columns to $375^{\circ}C$ for many months with little deterioration in resolution.

Wax Chromatography - The 80's Crossroads

The use of helium carrier gas for high temperature separations had little affect on the alkane resolution, and seriously distorted the response factors between C_{20} and C_{30} alkanes into a hump. The injection technique was investigated by carrying out tests with C_{17} to C_{44} alkane series standard. The syringe needle was retained in the injector for different lengths of time from zero to thirty seconds, and the best relative response factor results were obtained with a seven second dwell time.

GRAPH SHOWING THE RELATIVE RESPONSE FACTORS OF ALKANE STANDARDS FOR DIFFERENT CARBON NUMBER DISTRIBUTION METHODS

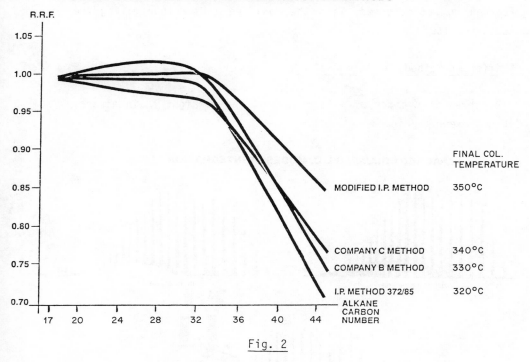

Fig. 2

The different carbon number distribution methods were tested using a C_{17} to C_{44} alkane series standard and by plotting the response factors for each alkane relative to C_{17} (see Fig. 2). For each method the relative response factors of the alkanes up to C_{32} are near to unity and then decline. The slope of the decline varies according to the G.L.C. method used and appears to be a function of the final column temperature. The best results were obtained using the Dussek Campbell Modified I.P. Method. An explanation could be that, under the test conditions, C_{40} elutes around $323°C$ and C_{44} around $343°C$. Therefore, with the other methods C_{40} and C_{44} elute isothermally and band broadening occurs.

The controlling factors in quantitative carbon number distribution analysis are:-

1. G.L.C. Method

   To test this a Standard Wax Blend (containing a standard mixture of paraffin and microcrystalline waxes) was analysed using each test method. Each method showed differences in resolution and "baseline hump" (the unresolved area underneath the alkane peaks). The higher the final column temperature of the method, the higher the carbon number detected, with the Modified I.P. Method giving the best results.

2. Different Methods of Integration

   The three main modes of integration of the straight chain alkanes can be seen in Fig. 3.

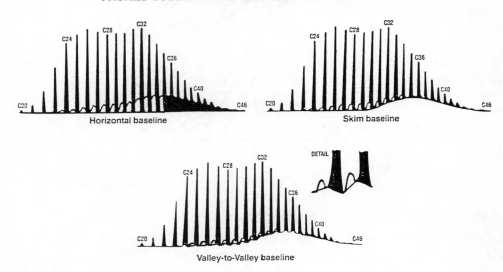

Fig. 3

To test the validity of the integration and calculation methods for straight chain alkanes, the Standard Wax Blend was injected and analysed each time by the Modified I.P. Method. The valley-to-valley integration results were very similar to the skim baseline results, therefore only the latter were plotted (Company B Method). The differences in results for the horizontal baseline (I.P. 372/85), and the skim baseline integration methods can be clearly seen in Fig. 4.

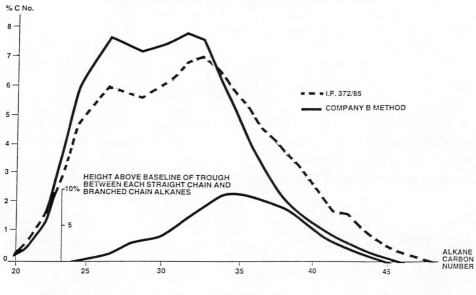

Fig. 4

The unintegrated "baseline hump" of the skim baseline method has been previously justified as being solely due to branched and cycloalkanes. However, the "baseline hump" distribution was found to be controlled by the chromatographic resolution. Also the validity of the skim baseline integration method (and the valley-to-valley method) are contradicted by the raw results (Fig. 5).

The amount of branched chain alkanes in the Standard Wax Blend was independently estimated (by molecular sieve analysis) to be approximately 25%. Fig. 5 shows the amount of straight chain (n-alkanes) for each integration method, using similar sized columns containing 5% Dexil 300 (top figure) and 5% OV-1 (bottom figure). The amount of branched chain alkanes indicated by the

RAW RESULTS FROM ANALYSIS OF STANDARD WAX BLEND BY DIFFERENT METHODS

Integration Method:- Calculation:-	I.P. 372/85 Horiz. Baseline R.R.F. Method	Modified I.P. Horiz. Baseline Internal Std.	Company B Skim Baseline Internal Std.	Company C V-V Baseline % Area only
% Total	85.1	80.1	62.4	97.2
n-Alkanes	83.8	78.3	—	—
Internal Std.	N/A	5.43	7.41	N/A
% Area	N/A	5.49	7.30	N/A

Fig. 5

valley-to-valley method is much too low (3%), whereas the skim baseline method is much too high (38%). The horizontal baseline method, indicating approximately 20% branch chain alkanes, appears to give the most feasible result.

3. The Maximum Column Temperature

$350^{\circ}C$ has been considered the normal maximum temperature for both OV-1 and Dexil 300 stationary phases coated onto packed columns. However, to test whether all of the alkanes had been eluted at this temperature, the standard wax blend was analysed for different maximum column temperature up to $400^{\circ}C$ using both types of column. The results are shown in Fig. 6 below:-

Maximum Column Temperature ($^{\circ}$C)	350	360	375	400
Maximum Detectable Carbon No.	47	50	54	54

Fig. 6

A maximum temperature of $375^{\circ}C$ appears to give the best results, with the unresolved hump of microcrystalline wax becoming apparent. Further the validity of the quantitative results using the Modified I.P. Method to $375^{\circ}C$ maximum column temperature, together with the horizontal baseline integration method, can be seen in the raw results below (Fig. 7).

As the column maximum temperature is increased to 375°C, the measured area for the C_{17} internal standard decreased towards the theoretical results of 5.0%. However, at this temperature, the figure for the amount at branched chain alkanes (approximately 16%) appears low. This is due to the lack of resolution of the branched chain alkanes after C_{38}. From the author's research, it would appear that the unresolved hump is caused by a mixture of straight, branched and cycloalkanes (with possibly some aromatics).

RAW RESULTS FOR STANDARD WAX BLEND ANALYSIS AT DIFFERENT MAXIMUM TEMPERATURES

TEMPERATURE PROGRAMME	150 to 350°C at 5°C min.$^{-1}$	150 to 360°C at 5°C min.$^{-1}$	150 to 375°C at 5°C min.$^{-1}$
Internal Std.	5.49	5.23	5.05
% Area	5.43	5.24	5.06
% Total	78.3	81.6	84.9
n-Alkanes	80.1	80.5	82.6

Fig. 7

Using carefully controlled conditions, it is possible to regularly use a 1.2m x 2mm i.d. packed column, containing 5% OV-1 or Dexil 300 stationary phase, up to 375°C to quantitively analyse waxes and wax blends. Also it is possible to programme the same columns up to 400°C for short periods, to separate high melting point microcrystallline wax to C_{72} (see Fig. 8), although the branched chain alkanes are unresolved. This method can also be used to separate Polywax 500 and 655.

Short, wide bore, thick film capillary columns have not as yet replaced packed columns as a rapid means of wax blend analysis, due to their maximum temperature being at present restricted to 330°C. However, medium length, narrow bore, thin film OV-1 type columns are excellent for separating waxes and wax blends (although analysis can take up to 50 minutes).

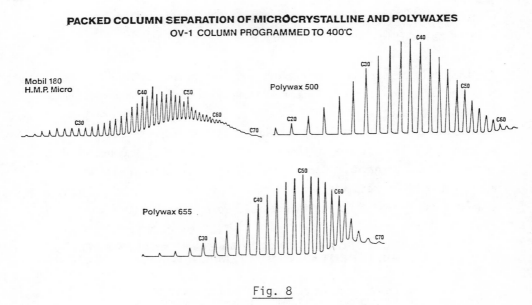

Fig. 8

In 1985 the author purchased a Perkin Elmer 8320 G.L.C. with a Programmable Temperature Vapouriser (P.T.V.) Injector. After a few initial problems, it was found that the P.T.V. heater could be operated constantly up to $450^{\circ}C$ without failing (as long as nitrogen was used, instead of compressed air, as the heating gas). Also it was found that the P.T.V. temperature had to be set $20^{\circ}C$ above the maximum column temperature so that all of the higher alkanes were completely eluted.

With the G.L.C. conditions optimised, it is possible to obtain a relative response factor of unity from C_{17} to C_{44}, even in the split mode. Different sizes of capillary column were used, in conjunction with several G.L.C. methods, to obtain the best separation of the standard wax blend. Fig. 9 shows that, although a 25m long standard OV-1 column produces a highly resolved chromatogram, it does not resolve the higher alkanes. Whereas, the 15m, thin film DB-1, (J&W) column programmed to $370^{\circ}C$ does and still retains the high resolution.

This column was found to give a reproducible quantitative separation of the standard wax blend, with most of the branched chain alkanes being resolved to give a total of approximately 30% (very similar to the molecular sieve result. Also the measured amount of C_{17} internal standard was very near to the theoretical result of 5.00%. It would

STANDARD WAX BLEND CAPILLARY COLUMN CHROMATOGRAMS

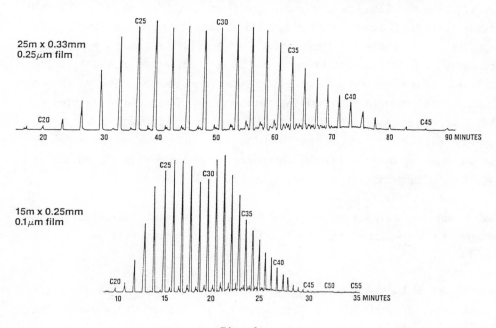

Fig. 9

appear that for both packed and capillary G.L.C. quantitative analysis of waxes and wax blends, most of the alkanes should be eluted before the maximum temperature. A comparison of the best packed column results with the DB-1 capillary column results showed a similar carbon number distribution, with the only differences being due to the lack of resolution of the packed columns above C_{35}.

A recent international interlaboratory exercise examined the capillary G.L.C. separation of paraffin waxes. The results of one laboratory using on-column injection onto a 15m DB-1 column, were very similar to the author's results using the same type of column and a P.T.V. injector. The author carried out an interlaboratory study, comparing different methods of injecting the Standard Wax Blend into the same size of OV-1 columns and temperature programming the columns to 350°C. A standard split/splitless injector could only resolve up to C_{48}, an on-column injector detected up to C_{53} and another laboratory using a P.T.V. injector managed to separate up to C_{51}. The author obtain similar results to the laboratory using the on-column injector, and managed to separate up to C_{55} by temperature programming the column to 375°C.

Recent arguments as to the relative merits of the on-column and the P.T.V. injector have been clouded by commercial interests. Several researchers have used biased comparative data to denigrate the P.T.V. injector. The author has shown that, if the P.T.V. injector is operated correctly, reliable and accurate quantitative separation of waxes can be achieved. Also the author has experienced problems in using on-column injection to separate high boiling point waxes. These are attributed by the author to condensation of the higher boiling fractions on the syringe wall during injection and to the injection valve becoming contaminated by higher alkanes, which are only eluted onto the column at high column temperatures (causing spurious final peaks).

The author has managed to separate high melting point microcrystalline wax up to $C_{70}+$ using the 15m DB-1 column programmed to $400^\circ C$ (see Fig.10).

The Polywax 500 and Polywax 655, were separated up to C_{80}, but a shorter column would probably produce better results. The polyimide coated 15m x 0.25mm i.d. DB-1 column has been constantly used up to $375^\circ C$ with only slow deterioration in the column resolution. This is probably due to the use of highly purified helium and the sealed carrier gas system.

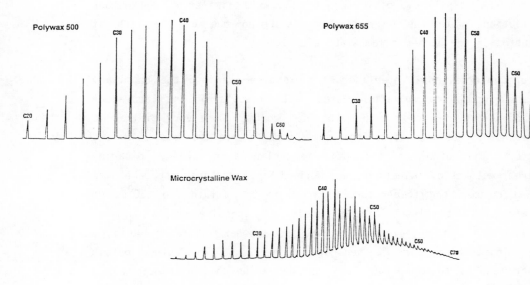

Fig. 10

Results have been published showing that the P.T.V. injector can be combined with short aluminium clad thin film capillary columns to elute a Polywax 1000 sample up to C_{120}, but the chromatogram lacks resolution. The author has started research into using Quadrex aluminium clad, narrow, thin film columns to endeavour to obtain total resolution of microcrystalline waxes to the highest carbon number.

The sole use of G.L.C. to investigate the analysis of waxes and polywaxes is questionable and probably leads to a biased perspective. Other methods, such as gel permeation chromatography (G.P.C.) are useful as a comparison. Prior to 1980, Hillman, et al, used several linked, heated columns of different pore sizes to separate waxes and polywaxes, without the aid of microprocessor technology.

The Standard Wax Blend and several Polywax samples, were analysed using two 20μ mixed porosity columns, an o-dichlorobenzene carrier solvent at 140°C and a refractive index detector. Accurate calibration was carried out using standards of similar hydrodynamic volume, i.e. polyethylene. This gave G.P.C. results for the mean molecular weight and the molecular weight spread of the Standard Wax Blend and the Polywax 500 standard very similar to the G.L.C. results. Recent advances in G.P.C. data handling should make this technique more accurate for molecular weight distribution analysis of microcrystalline and polywaxes.

Supercritical Fluid Chromatography (combined with F.I.D.), overcomes the volatility constraints of G.L.C., and theoretically combines the best features of G.L.C. and G.P.C. By using highly compressed carbon dioxide plus temperature/pressure programming, it is possible to separate out high molecular weight waxes on a short narrow capillary column coated with OV-1 (see Fig. 11).

However, there are several problems, including the complexity of the analytical method. Pressure programming causes uneven resolution, lower than that obtained by high temperature capillary G.L.C. There are also problems associated with the solubility of the high concentrations of microcrystalline, or polywaxes required by the split injection valve used, which leads to biased quantitative results. This problem is now being overcome by using heated loop injection and a rapidly rotating valve, which gives a minute volume on-column injection.

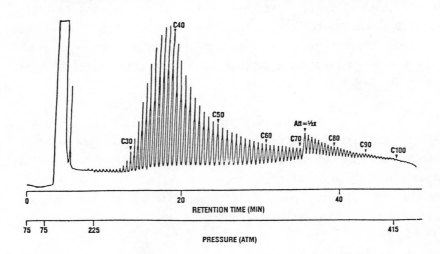

Fig. 11

It is also possible to use other techniques to elucidate the meaningfulness of high temperature G.L.C. of waxes. An accurate mean molecular weight can be obtained using modern ebullioscopic techniques; differential thermogravimetry can be used to produce a simulated distillation curve; and a 5 Å molecular sieve can be used to measure the total branched chain alkanes.

In conclusion, it should be noted that packed column G.L.C., correctly used to elute all alkanes, is still relevant for quality control of waxes. It is also important, in both packed and capillary G.L.C. quantitative analysis, that an internal standard (e.g. C_{17}) and a horizontal baseline integration method are used. Wide bore, thick film capillary columns cannot, at present, supersede packed column G.L.C. for rapid Q.C. analysis of waxes, and more work is required in this area. Supercritical fluid chromatography can be used to produce a comparative carbon number distribution, but it is still a complex analytical technique giving lower resolution than high temperature capillary G.L.C. More comparative research is required to obtain an unbiased perspective of high temperature G.L.C. and the different methods used. The final goal is to obtain highly resolved chromatograms (even for microcrystalline wax), so that the raw material results can be computerised to accurately predict the composition of wax blends and thus save production time.

I wish to thank all those who helped in my research and the Directors of Dussek Campbell Ltd. who made it possible.

Ref: A.D. Barker M.Sc Thesis - January 1987

Thermal Analysis—I.P. Studies

S.R. Wallis
Castrol Research Laboratories, Pangbourne, Reading RG8 7QR

Five years ago, the Standardisation Committee of the Institute of Petroleum (I.P) established a panel (ST-G-9) concerned with thermal analysis (TA) techniques[1]. The panel has been enthusiastically supported and has representation from oil and additive companies, engine manufacturers, a university, makers of TA equipment and lube users, such as the Ministry of Defence and public transport organisations.

The general aim is to produce standard methods of analysis or testing which are acceptable to the oil and related industries. TA is now more widely used in these industries and ST-G-9 aims to produce TA methods that are compatible, as far as possible, with most makes of instrument.

The TA techniques considered by ST-G-9 are thermogravimetry (TG), differential thermal analysis (DTA) and differential scanning calorimetry (DSC). Two proposed methods have recently been submitted for inclusion in the 1988 I.P Methods book, one using TG, the other DTA or DSC, and this paper describes them.

WAX APPEARANCE TEMPERATURE (WAT) OF MIDDLE DISTILLATE FUELS BY DIFFERENTIAL THERMAL ANALYSIS (DTA) OR DIFFERENTIAL SCANNING CALORIMETRY (DSC)

The scope of the method covers the determination of the temperature at which waxy solids form when middle distillate fuels are cooled. The term "middle distillates", includes gas oils, burner fuels and diesel fuels, but not jet fuels (aviation turbine kerosines).

DTA and DSC are techniques in which the temperature difference or heat change between a substance and a reference is measured as a function of temperature whilst the substance and reference are subjected to a controlled temperature programme. In the present context, the DTA and DSC techniques are regarded as identical.

Figure 1A shows a schematic DSC trace illustrating the type of features that can be observed. The shift in the baseline, due, for example to a glass transition temperature or change in the specific heat of the sample are useful effects but are not concerned with this paper. The negative endothermic peak is typical of a melting point and, conversely, the positive, exothermic peak is typical of a freezing point. There are other chemical and physical reasons for endotherms and exotherms (for example, oxidation processes yield strong exotherms), but this

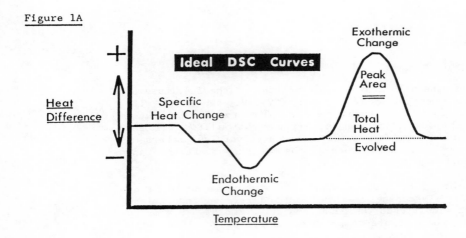

Figure 1A

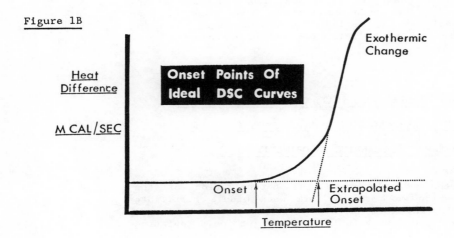

Figure 1B

Figure 1   A (upper). Schematic of DSC trace showing the type of features observed.

B (lower). Schematic DSC trace (expanded), showing the measurement of extrapolated onset temperature.

WAT method is concerned only with melting and crystallisation (freezing). Part of the exotherm feature is expanded in Figure 1B to illustrate the measurement of <u>extrapolated onset temperature</u> (which can be measured for an endothermic peak as well as an exotherm).

The significance of the method is that the WAT of a fuel is related to the lowest temperature of its utility for certain applications. For example, if wax begins to crystallise from diesel fuel in cold weather, the viscosity of the fuel increases markedly, making engines difficult to start and risking blockage of filters etc. The method requires only a very small sample and is quick: if a laboratory is set up for maximum sample through-put, one run takes about ten minutes.

In summary; the fuel is cooled at a constant rate of 10°C/minute with a DTA or DSC instrument and the onset temperature of the crystallisation exotherm of the wax in the sample is recorded. The work was done on instruments manufactured by Du Pont, Mettler, Perkin-Elmer and Stanton-Redcroft.

The instruments are calibrated for temperature (Figure 2) using the melting points of pure (99.999%) mercury (-38.9°C) and indium (156.6°C). Flat-based aluminium sample pans are satisfactory as appropriate to the make of instrument: work has shown that it is immaterial whether the pans are surface treated or untreated aluminium (in reality, an aluminium oxide surface). 10 microlitres of sample yields adequate sensitivity and does not overfill the pan. The pan is then sealed with an aluminium lid using a crimping press. Pans and lids are solvent washed and dried prior to use.

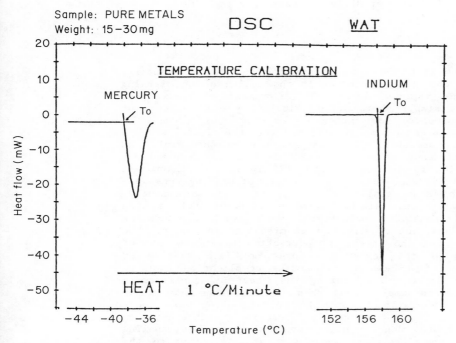

Figure 2   Temperature calibration: WAT method

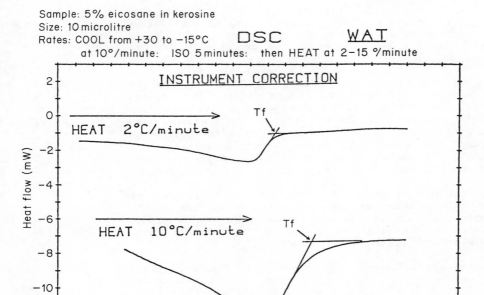

Figure 3 Final melting point Tf of eicosane (5%) in kerosine at different heating rates.

The WAT varies with the rate of cooling due to two factors: supercooling, and the differences in the design of the sensing heads of the various makes of instrument. Supercooling variations are simply obviated by stipulating a fixed rate of cooling; 10°C/minute was chosen as optimum. Differences in instrument design can result in variations in the time at which a heat change is observed. This is overcome by determining an empirical instrument correction (expressed as a temperature) which is applied to all experimental WAT runs. The instrument correction (C) is determined using a series of heating runs on a standard which is of similar nature to the samples; namely, eicosane, a waxy, pure alkane dissolved (approximately 5%) in kerosine. Alternatives to kerosine were examined, such as heptane or decane, but the traces were misshapen and noisy. The numerical value of the instrument correction is independent of the direction of the temperature programme (i.e heating or cooling); use of a heating run avoids complications due to supercooling. In the WAT cooling run, heat is continuously abstracted from the sample, which then suddenly emits heat as the wax begins to crystallise. With a finite cooling rate, heat emission is sensed at a fractionally later time (which equates to a lower temperature) than at an infinitely slow, or zero, cooling rate. Hence, the instrument correction (C) is <u>added</u> to the experimental onset temperature. The instrument correction needs

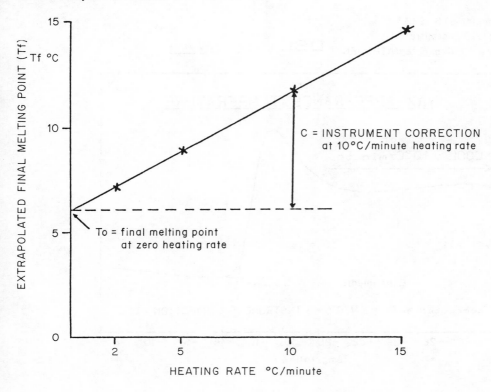

Figure 4   Plot of final melting point Tf of eicosane (5%) in Kerosine against heating rate.

to be determined only infrequently (say every two months) unless some change has occurred to the sensing head. The eicosane/kerosine standard is heated from - 20°C to 20°C at, say, four rates (for example; 2, 5, 10, 15°C/minute) and the final melting points Tf recorded, Figure 3. Tf is plotted against heating rate and the difference (C) between the intercept temperature and Tf at the required programme rate, 10°C/minute, is the instrument correction (C). Figure 4.

A typical WAT trace is illustrated in Figure 5. The sample is cooled from + 30°C to - 40°C; the experimental onset of wax crystallisation W°C is observed on the trace at - 15°C, and then the instrument correction is added, such that:

corrected WAT = W°C + C

Results on three of the round-robin samples from five laboratories are listed in Table 1. When the statistics of all collaborative test results have been assessed we expect a repeatability between 1 and 2°C and a reproducibility of 4°C This precision is intermediate between the two other common methods of assessing wax precipitation, Wax Appearance Point (WAP, ASTM D3117) and Cloud Point (CP, IP219/ASTM 2500).

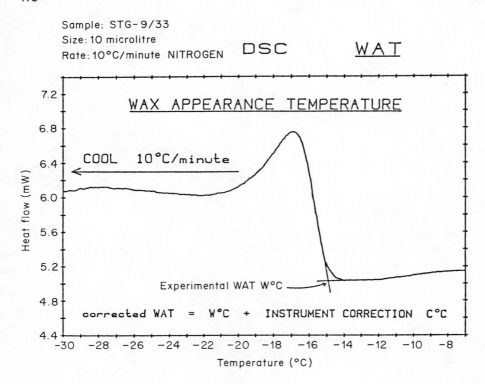

Figure 5   Typical WAT DSC trace.

Table 1.

WAT TYPICAL RESULTS

Expected Precision : repeatability 2°C; reproducibility 4°C

LABORATORY				
1	2	4	5	6
+5.5, +5.8	+4.4, +4.5	+4.7, +4.7	+5.9, +5.7	+5.3, +4.8
−4.9, −3.7	−4.8, −7.8	−5.2, −5.0	−4.1, −3.8	−3.9, −4.0
−9.9, −10.4	−13.8, −13.9	−12.9, −12.9	−10.6, −10.6	−11.0, −10.8

The evidence indicates that the WAT should be interpreted in the same way as the more traditional WAP and CP and that there is a strong correlation between them. One laboratory represented on the Panel has used the WAT method for several years. Work from this laboratory plotted in Figure 6 shows WAT by DSC of 25 base fuels plotted against the corresponding WAP and CP values. Although there is obvious scatter, a believable 1:1 correlation is apparent. Recent work in the literature[2] on nearly one hundred gas oil, kerosine and diesel oil samples, using a similar DSC method has shown a very close correlation (correlation coefficient 0.998) between the DSC and CP methods.

For maximum sample through-put it is necessary to fit a mechanical cooling accessory to the calorimeter head. For occasional use, cooling by means of a liquid nitrogen canister is satisfactory but much less convenient and more time -consuming.

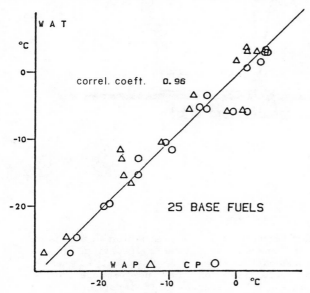

Figure 6   Correlation of WAT with Wax Appearance Point and Cloud Point of base fuels.

VOLATILITY OF AUTOMOTIVE LUBRICATING OILS BY

THERMOGRAVIMETRY (TG)

TG is a technique which measures the change in weight (mass) of a sample as a function of temperature as the sample is subjected to a controlled temperature programme. Figure 7 is a schematic TG trace which illustrates the features. In the present context of volatility, the definition can be made more specific:- TG measures the <u>loss</u> in weight (almost always expressed as a percentage of the initial weight) as the sample is subjected to a controlled temperature <u>rise</u>. The derivative of the TG trace can give useful information in many circumstances, but it was not found helpful in the present work.

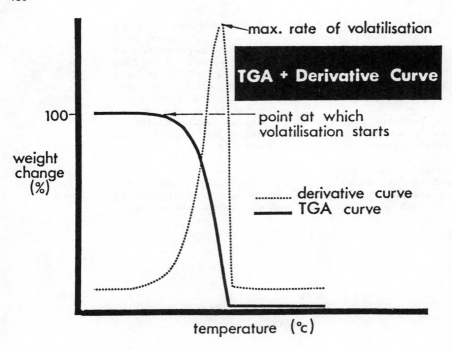

Figure 7   Schematic TG trace.

The scope of the method covers the determination of the volatility of base oils and unused formulated lubricating oils for diesel and gasoline automotive engines. Volatility is an important characteristic relating to the performance of automotive lubricating oils, and has increased in importance in recent years due to the trend towards thinner oils used to promote greater fuel economy. Available methods (such as gas chromatographic or distillation methods) are restricted to the more volatile oils, or, with the various isothermal methods (such as the Noack Test, DIN 51581) the information is limited to the weight loss at a fixed temperature. This proposed TG method yields a profile of the weight loss/temperature relationship; it provides a continuous examination of weight loss with time as temperature rises.

In summary, the oil is heated in a thermobalance at a constant rate of 10°C/minute from 40°C to 550°C under flowing nitrogen. The weight loss (%) as a function of temperature is recorded. A typical trace is shown in Figure 8. As an alternative, the result can be presented in tabular form, by listing the temperatures corresponding to particular weight losses (also shown in Figure 8). The tabular presentation is useful when interest is focused on the volatility of a particular oil; but, commonly, a comparison is being made between more than one oil, and it is in this circumstance that the benefit of visual comparison of the volatility curves (by overlaying) is apparent.

Thermal Analysis - I.P. Studies

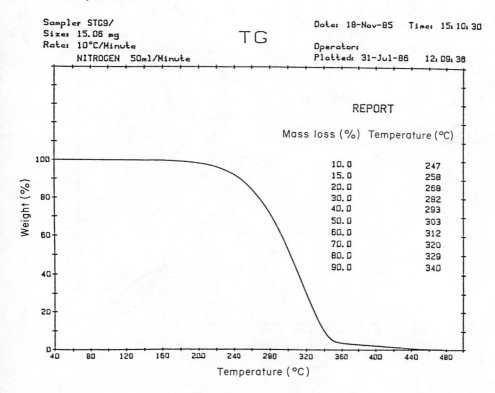

Figure 8   Typical Volatility TG trace.

For example, in Figure 9, it is immediately clear that the two oils displayed have different volatility profiles. Furthermore, the profiles cross each other, such that the oil of greatest volatility at the lower temperatures exhibits least volatility at higher temperatures. This crossover effect is not uncommon and is particularly important when it occurs within, say, the initial 20% of weight loss. In some cases, additional information is more readily appreciated from the profile than from the tabular presentation. One of the traces in Figure 9 shows a double hump indicating that the lubricant was constituted from base fluids of widely different volatilities ("dumb-bell blend").

The two profiles shown in Figure 9 also illustrate visually the repeatability to be expected from the method: both profiles are constructed from duplicate runs (solid and dashed lines). The near superimposition of the double-humped trace illustrates a level of repeatability that is commonly achieved. The small difference between the duplicate runs of the other profile is still acceptable, but bordering towards the unacceptable. A likely cause of poor repeatability is condensation on the balance supports etc. Due to this, the method states that the TG instrument must be purged at 800°C under air or oxygen for 15 minutes at the completion of each day's runs, or more frequently if condensation is suspected.

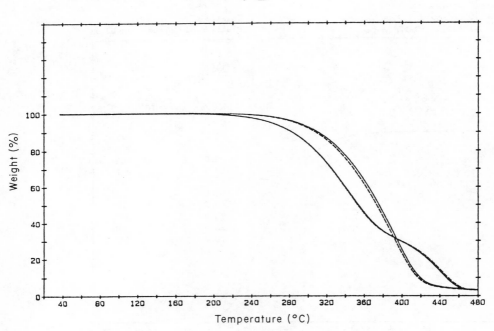

Figure 9   Comparison of the volatility of two oils: duplicate TG determinations on both oils.

The round robin work for the proposed method was carried out on instruments manufactured by Du Pont, Mettler, Perkin-Elmer and Stanton-Redcroft. Apart from the proviso (condensation) above, satisfactory short-term repeatability was achieved, but the attainment of acceptable reproducibility or long-term repeatability required close attention to two aspects: temperature calibration, and pan geometry.

Temperature calibration of TG instruments is complicated by the fact that the sample is not in contact with the thermocouple (except for simultaneous TG/DTA instruments). Two techniques are now available: the Curie Point method[3-6] and the metal melt method[7-10]. In the Curie Point method, a permanent magnet is positioned round the furnace and a Curie Point alloy is used as the sample. The alloy undergoes a change in its magnetic properties at the Curie Point temperature and, due to the influence of the magnetic field, this is registered as an apparent weight change by the TG instrument. The metal melt method has two variations (the action/reaction method and the melt drop method) but both depend on the melting point of a pure metal in the form of wire. In the action/reaction method, the wire is arranged such that, on melting, it falls into the pan, causing the balance to "bounce" and record a sharp incursion of the trace at the melting point. In the melt drop method, the wire is formed into a loop from which is suspended a 50mg weight;

when the metal wire melts, the weight drops through a hole in the pan and a sharp weight loss is recorded at the melting point. All the methods are acceptable; see references for details. The proposed volatility method does not give the details since it is regarded as part of the expertise of an experienced TG operator and also it would make the method much longer. The proposed method merely refers the reader to the relevant ASTM draft method[11] (TM-01-03-B-4), which provides full details, in the expectation that the draft will be published in the near future.

On the whole, the ST-G-9 panel prefers the metal melt method since there is debate with the Curie point method about which portion of the trace (onset, midpoint, or end) to take as the calibration temperature. It is emphasised that to obtain reproducible results, the TG instrument must be calibrated regularly, say twice per week if in continuous use, and certainly after any disturbance to the system (such as movement of the thermocouple relative to the sample pan).

Pan Geometry is the other aspect of the technique which required close attention in order to obtain reproducible results. Figure 10 shows sketches of typical TG sample pans; they differ in diameter, thickness, shape and depth. The effect these differences have on the volatility temperature is illustrated in Table 2 where the temperature corresponding to 10% weight loss varies by up to 24°C.

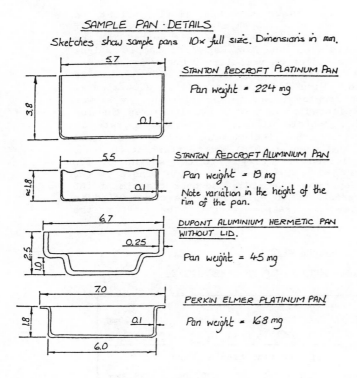

Figure 10   Sketches of typical TG sample pans.

Table 2   TG VARIATION WITH PAN TYPE

	PAN TYPE (20 mg sample) as sketched in Figure 10			
	S-R 4mm Pt	S-R 2mm Al	DuP hermetic Al	P-E Pt
Temperature °C at 10% weight loss	271	254	255	247

It was necessary to place a tight specification on the sample pan to get acceptable reproducibility. In practice this meant specifying the exclusive use of a particular aluminium pan provided by one manufacturer. The choice was based upon the smallest available pan such that it fitted all the makes of instrument used.

Three other parameters had to be considered and optimised: nitrogen flow rate, heating rate and sample weight.

Surprisingly, the nitrogen flow rate had little effect on volatility within the range 25 - 100 ml/minute. A flow rate of 50 ml/minute was chosen for the method since this suited all the makes of instrument examined. Heating Rate and sample weight variations exert a large effect on volatility, but these did not cause problems once set to the optimum values. A heating rate of 10°C/minute was chosen. With a faster rate (e.g. 20°C/minute) discrimination between samples began to deteriorate; a slower rate (e.g 5°C/minute) offered no advantage but increased the run time. The effect of a tenfold difference in heating rate is shown in Figure 11. The method includes an initial isothermal equilibration period (5 minutes at 40°C) at the start of the run. Sample weight was optimised at 15 mg. If the weight is too small, then sensitivity is impaired; if too large, there is the risk of the sample creeping up the sides and over the rim of the pan. Sample weight variations on volatility are exemplified in Table 3: the greater the sample weight, the higher the

Table 3.   SAMPLE WEIGHT

GREATER SAMPLE WEIGHT ≡ HIGHER TEMPERATURES

Temperature °C at 20% weight loss	Stanton-Redcroft 2 mm Pt pans			
	2mg	5mg	10mg	20mg
	221	248	258	271
	Du Pont flat base Al pans			
	2.5mg 5.1mg	11.1mg	20.1mg	29.8mg
	226 238	250	263	271

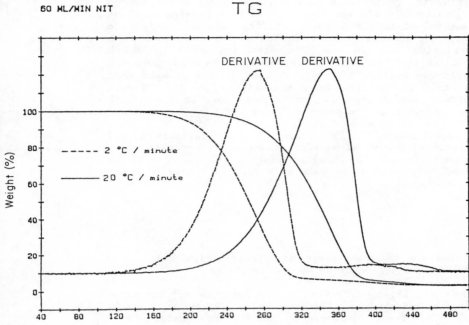

Figure 11   Effect of heating rate on volatility.

volatility temperature. A tenfold increase in sample weight will raise the volatility temperature at a particular weight loss by 40 to 50°C. Hence, a tight limit of ± 0.1 mg is applied to the sample weight.

Some typical results are listed in Table 4. When all the round-robin results have been examined statistically we are expecting a repeatability of about 3°C and a reproducibility about 6°C.

Table 4.   VOLATILITY BY TG - TYPICAL RESULTS

WEIGHT LOSS %	TEMPERATURE °C at WEIGHT LOSS				
	STG9/29			STG9/48	
	LAB 1		LAB 2	LAB 1		LAB 2
10	249 , 248	252	249 , 247	248
15	261 , 261	264	261 , 258	262
20	271 , 270	276	269 , 267	269
30	285 , 284	287	281 , 281	282
40	296 , 296	297	291 , 291	290
50	306 , 306	307	299 , 299	297
60	315 , 315	317	306 , 307	305
70	324 , 324	323	312 , 314	309
80	333 , 333	333	318 , 321	316
90	344 , 344	341	325 , 328	321

One popular method used for assessing volatility is the Noack Test.[12-13] The ST-G-9 panel used some oils of known Noack volatility for its round-robin work and consequently is able to state that, based on a limited amount of work, the weight loss at 260°C is equivalent to the Noack volatility within fairly wide limits (awaiting statistics, but probably $\pm$ 1% absolute). A close correlation between the TG volatility and the Noack volatility is unlikely to be justified because the conditions for the two tests are quite different. However it was thought that extraction of an appropriate Noack volatility from the TG data could provide a useful screening value so a note on the "Noack Temperature" is given in the proposed TG method.

The ST-G-9 panel think it likely that the TG volatility method can be extended with minimal modification to most types of oils, including used oil and the heavier fuels. The panel will be shortly commencing round-robin work on a variety of sample types to increase the scope.

Acknowledgements

It is a pleasure to acknowledge the laboratories which have participated in the I.P panel on Thermal Analysis:

Associated Octel, Ellesmere Port; Austin Rover Group, Coventry; B.P. Research, Sunbury; British Rail, Derby; Castrol, Pangbourne; Du Pont, Stevenage; Ethyl, Bracknell; Esso Research, Abingdon; London Regional Transport, Chiswick; Lubrizol, Derby; Mettler, Crawley; Mobil, Coryton; MoD Directorate of Quality Assurance/Technical Support, Harefield and Woolwich; Perkin-Elmer, Beaconsfield; Perkins Engines, Peterborough; Shell Research, Thornton; Stanton-Redcroft, Wimbledon; University of Leeds. Also, communications with Subcommittee E37-01 of the ASTM are acknowledged. The WAT method originated from Exxon and Esso laboratories.
The support of Castrol Ltd. in the preparation of this paper is also gratefully acknowledged.

REFERENCES

1. Wallis S. R., Analytical Proceedings, 1985, 22, 44

2. E-L. Heino, Thermochimica Acta, 1987, 114, 125

3. S. D. Norem, M. J. O'Neill and A. P. Gray, Thermochimica Acta, 1970, 1, 29

4. R. L. Blaine and P. G. Fair, Thermochimica Acta, 1983, 67, 233.

5. P. D. Garn and A-A. Alamolhoda, Analytical Calorimetry, 1984, 5, 33.

6. P. K. Gallagher and E. M. Gyorgy, Thermochimica Acta, 1986, 109, 193.

7. A. R. McGhie, Analytical Chemistry, 1983, 55, 987.

8. A. R. McGhie, R. L. Blaine, J. Chiu and P. G. Fair, Thermochimica Acta, 1983, 67, 241.

9. A. A. J. Cash, P. G. Laye and A. L. Panting, Thermochimica Acta, 1987, 113, 205.

10. E. L. Charsley, S. St. J. Warne and S. B. Warrington, Thermochimica Acta, 1987, 114, 53.

11. ASTM TM-01-03-B-4: Standard Practice for Calibration of Temperature Scale for Thermogravimetry; draft method in preparation.

12. German Standard, DIN 51581; Determination of Evaporation Loss of Lubricating Oils (Noack Method), 1977.

13. CEC/ELTC CL-34 Project Group, MoD Defence Standard 05-50/1: Evaporation Loss of Lubricating Oils by the Noack Method; draft unified method, in preparation.

The Use of Inductively Coupled Plasma Emission Spectroscopy (ICES) in Residual and Distillate Fuel Analysis

B. Pahlavanpour* and E.K. Johnson
* Caleb Brett Laboratories, Kingston Road, Leatherhead KT22 7LZ
77 Stockbridge Road, Winchester, Hampshire, SO22 6RP

SUMMARY

The development and use of Inductively Coupled Plasma for optical emission spectrometry is described. The use of ICPES for routine analysis of distillate and residual fuel oil samples enables rapid, accurate and simultaneous determination without interference of up to 15 metallic elements present including copper, zinc, sodium, lead, nickel, barium, potassium, iron, manganese, phosphorus, silicon, aluminium, vanadium, calcium, and antimony.
[Cu Zn Na Pb Ni Ba K Fe Mn P Si Al V Ca Sb.]

The paper includes comparisons with supporting tables
A. 2 fuel samples tested for 15 elements using 3 digestion methods. Zn,Na,Pb,Ni,Ba,K,Fe,Mn,P,Si,Al,V,Ca,Sb,Cu.
B. 6 fuel samples tested at 2 laboratories for 13 metals Zn, Pb, Ba, K, Fe, Mn, P, Si, Al, V, Ca, Sb, Cu.
C. Si,Al & Na determined in 2 laboratories with ICPES and AAS using metaborate fusion and the method of IP 363
D. Analysis on same sample in septuplicate for 12 elements with indication of averages and repeatability.
E. 6 element analysis by ICP compared with Colorometric/AAS

The main advantages of this ICPES method are :-
  accurate analysis can be carried out at reduced cost,
  the method requires minimum operator involvement,
  small amounts of contaminants are added with the reagents.

The availability of rapid accurate data on the presence of metal elements in fuel oils enables diagnosis of both natural constituents and contaminant levels of critical elements. As a result advice can be given to the users of the fuel directed to optimising use of the fuel or minimising the adverse consequences of using a particular fuel.

1. INTRODUCTION

Conventional techniques for elemental analysis, being colorimetric, rely on complexing the metals with reagents and measuring the absorption of colour complexes. These techniques are time consuming, are subject to interference and are only available for a limited number of elements such as Copper, Vanadium, Aluminium and Silicon.

The common techniques for trace element analysis, are atomic absorption spectroscopy (AAS) and inductively coupled plasma atomic emission spectroscopy (ICPES). Many elements are difficult or impractical to determine by A.A.S. and different procedures are required for specific elements By comparison, inductively coupled plasma-atomic emission spectroscopy can provide simultaneous information on up to 25 elements of interest with a degree of precision which is available only from the steady-state signals of ICPES.

2. USE OF ICPES FOR TRACE ELEMENT ANALYSIS.

The development of the Inductively Coupled Plasma for optical emission spectrometry has provided the analytical chemist with a powerful technique for elemental determination in a variety of sample types. (1,2)

The use of inductively coupled plasma emission spectrometry (ICP) in fuel analysis is rapidly increasing and the method is likely to become predominant in elemental analysis generally. The potential uses of a rapid cost-effective method of multi-element analysis have long been apparent to workers in oil analysis.

Elemental analysis on ashed fuel, including the determination of Silicon, is performed with solutions obtained from fusions, although in some experiments and routine analyses (3,4), samples are diluted with an organic solvent, such as xylene, and sprayed directly into the ICP. However, this technique is not suitable for fuel oil samples due to presence of catalytic fines which are likely to block the nebulizer. Therefore present quantitative fueloil analysis using spectrometry has to be a solution technique.

A number of reagents have been used as fluxes in the preparation of solutions for analysis from ashed materials. No attempt has been made to document or evaluate all possible reagents and for methods, but certain trends may be noticed. Some reagents, although excellent for single element determinations on ashes, are useless for multi-element analysis (5). The most suitable reagents for a particular application is one which is pure with respect to the trace element(s) to be determined.

Classical methods of analysis relied heavily on the use of sodium carbonate as a flux. This procedure could be used for ICP analysis, but is unlikely to achieve widespread popularity because it excludes the possibility of determining sodium.

In recent years borate fluxes have been widely used; lithium tetraborate, in particular, has been used successfully to determine most of the major elements (6). Lithium tetraborate is also well suited for ICP analysis as a relatively low flux-to-sample ratio of 1:1 can be used.

Most ICP systems do not operate well when the solid content of the solution exceeds 2%. Solutions obtained from fluxes used with a 10:1 ratio with the sample require considerable dilution before they can be analysed by ICP.

This paper describes a method for the determination of silicon, iron, aluminium, sodium, vanadium, calcium, nickel manganese, and lead in fuel that gives results comparable with those obtained by acid extraction and sodium carbonate fustion, but can be adapted to the rapid analysis of large batches of samples.

3. INDUCTIVELY COUPLED PLASMA EMISSION SPECTROMETRY.

In this section the main features of ICPES are given.

3.1 ICP excitation source

In the use of an emission spectrometer for elemental analysis an excitation source requires the following:-
  (i) capability of exciting lines of many elements;
  (ii) high sensitivity and low background;
  (iii) linearity of calibration;
  (iv) freedom from interferences;
  (v) convenience of operation;
  (vi) good stability.

In flame sources the temperature of the gas is limited by the heat of combustion of the fuel. No fundamental limitation such as that is present with electric discharges and factors such as the excitation capability, sensitivity, and freedom from chemical interference are improved with an increased temperature. This has led to the development of plasma sources.

A plasma is a gas ionized to a significant degree. An important property of a plasma is that large quantities of electrical energy can be transferred to it if it is sufficienctly ionized, and this supply of energy can heat it to very high temperatures.

A simple scheme for the use of such a spectroscopic source is to pass the sample into the very hot region to ensure complete dissociation to the atoms, and use the cooler part of the plasma for emission. Such a reduction in temperature may also allow the sample atoms to reform molecules but usually dissociation energies are less than ionization energies, so that when the sample is diluted by the plasma gas, a position in the tail flame may be found where there has been no significant molecular association but where the continuum is weak.

In the high frequency inductively-coupled plasma, a stream of ionised argon gas is contained in a circular quartz tube surrounded by a work coil carrying high-frequency alternating electric current. The argon is heated by induction in the high frequency field. The quartz tube is cooled by a second stream of argon. The high frequency plasma is similar to a transformer, the work coil being the primary and the ionised gas a one-turn secondary.

Greenfield (7) reported that as the temperature was increased in plasma gas the emission intensity increased. However the actual temperature depended on the torch geometry, the gas flows and the power supplied, The gas carrying the nebulised sample droplets flows through the cooler central tunnel of the plasma which is surropunded by the very hot toroidal region where the applied power is dissipated. When this gas flow emerges from the tunnel, it forms a well-defined and narrow tail-flame. Because of its small radius, the emission of the analyte atoms per unit of surface area is high, and this effect contributes to the sensitivity.

The continuum background is low in the tail flame and the combination of high sensitivity and low noise on the background gives good detection limits. The formation of refractory compounds in flame methods effectively locks up atoms and prevents them from contributing to the atom reservoir, thus reducing the emission intensity. This formation of refractory compounds is not observed in ICPES, and consequently there is a greater degree of freedom from chemical interference.

ICPES is relatively free from self-absorption because the analyte atoms are confined to the tunnel and the narrow tail-flame and there are very few atoms in the cooler regions that surround the tail-flame. Also because of low background noise, a wide linear calibration range is obtained.

Greenfield et al. (1975) (8) reported that the dimensions of the central injector tube are critical, but any change in other dimensions could be compensated by changing gas flows.

For plasma stability, the most important factor is concentricity of the work-coil and the torch. The torch was made of three concentric tubes, the two outer ones being made from silica glass contain the plasma, and the inner tube of borosilicate glass is for injection of the nebulised sample by carrier gas. The plasma gas was argon and the coolant gas nitrogen. Various forms of pneumatic nebulizer were tried. Ultra-sonic nebulization was found to be useful only for single samples and after each sample the tube had to be washed.

In 1965 Wendt and Fassel (9) employed various experimental configurations for formation of plasmas of the inductively coupled type. An ultrasonic atomizing system similar in design to that of West and Hume (10) was used to introduce the sample solution into the plasma. The torch was similar in design to Greenfield's but of smaller diameter.

Boumans and DeBoer (11) studied an experimental selection of conditions under which a low power inductively coupled plasma can be operated for simultaneous analysis of many elements. The plasma torch was of the now well-established type with three concentric tubes, but with only two gas flows: the coolant gas through the annular cavity between outer and intermediate tube and the carrier gas flow through the central injector tube. They examined the effects of varying the geometry of the torches, the effect of carrier gas flow rates, solution uptake rates and the size of the fog chamber.

Summarising ICPES is an analytical method of great potential in fuel analysis and applied geochemnistry, combining the virtues of
* low detection limits
* freedom from interferences,
* linear calibration over a wide range of concentrations
* simultaneous multi-element capability.

4. EXPERIMENTAL DETAILS

4.1. Apparatus
  1. Plasma system.
     The spectrometer was an Applied Research Laboratories ARL 34000 Quantorvac 1-m vacuum spectrometer.
     The wavelengths used are listed in Table 1.
     The details of the instrumentation and conditions are given Table 2.
  2. Platinum basin, 100 ml capacity, cleaned with fused potassium hydrogen sulphate.
  3. Disperser (homogenizer) - to achieve complete homogeneity of sample.
  4. Electric magnetic stirrer hot plates, capable of maintaining a temperature of 70°C.

4.2. Reagents
  All reagents were of analytical grade.
  1. Hydrochloric acid, - concentrated.
  2. Propan-2-ol (isopropyl alcohol).
  3. Toluene
  4. Stock solutions containing 1000 ug/ml of each ion.
  5. Potassium Hydrogen Sulphate, fused.
  6. Tartaric Acid.
  7. Aqueous solution containing 0.5% wt/vol tartaric acid and 3% vol/vol hydrochloric acid.
  8. Flux - spectroflux type 120 A was obtained from Johnson Matthey Chemicals. This flux comprises of 90% lithium tetraborate and 10% lithium flouride. (lithium fluoride is necessary to prevent heavy metal corrosion of the platinum crucible and to lower the fusion temperature).

5. EXTRACTION / DIGESTION PROCEDURES

Analytical methods for metallic elements in fuel oils require decomposition of the sample and the liberation of the analyte elements into solution before their concentrations can be determined.

However, when organic matter has to be destroyed as a preliminary to the determination of metallic traces, the choice of decomposition method is determined by both

(a) the nature of the organic material and of any inorganic constituent and
(b) the method of determination for the particular metals.

The tetraborate fusion technique can be applied with ICP spectrometry for the determination of up to 21 elements in an ashed fuel oil. The metallic elements are extracted from the ash by fusion with lithium tetraborate, and the fused material is dissolved in a 3% hydrochloric acid and 0.5% tartaric acid solution.

The ashing and fusion is usually completed overnight. Fusion and dissolution occupies approximately 1/2 hour. The spectroscopic procedure, including integration of the spectrum to give a calculated result on all 21 elements, can be completed in less than 2 minutes.

The tetraborate fusion procedure developed to give economic use of tetraborate with minimum interference from any extraneous elements will now be described.

5.1. Ashing and Tetraborate fusion. (Method A)
1. Weigh a clean platinum basin to the nearest 0.1 mg.
2. Place the sample container in an oven at a temperature between 50°C and 60°C and maintain the sample at this temperature until all the sample has melted and reached a uniform viscosity. Insert the shaft of a high speed homogenizer into the sample container so that the head of the shaft is immersed to approximately 5 mm from the bottom of the sample vessel. Homogenize the sample for about five minutes.
3. Immediately transfer up to 20g (but not less than 15g) of the well mixed sample, to the platinum basin and re-weigh the basin and contents to the nearest 0.1 g to obtain the weight of sample taken.
4. Warm the basin and contents gently with a bunsen flame until the sample can be ignited. If the sample contains considerable amounts of moisture, foaming and frothing may cause loss of sample. If this is the case, discard the sample and to a fresh portion add 1-2 ml of propan-2-ol before heating. If this is not satisfactory add 10 ml of a mixture of equal parts of toluene and propan-2-ol and mix the sample thoroughly. Place several strips of ashless filter paper in the mixture and warm gently. When the paper begins to burn, the greater part of the water will have been removed.
5. Maintain the contents of the basin at a temperature such that all the combustible material is removed and only carbon and ash remain.
6. Place the basin and contents in a muffle furnace maintained at a temperature between 525 and 575°C.
7. Maintain the muffle furnace at the above temperature until all the carbon has been removed and only ash remains Normally this will take 3 to 4 hours but it can be left overnight).
8. Cool the basin to room temperature weigh the basin to obtain the weight of ash. If the weight of ash is less than 0.1 g. add 0.2 g. of flux to the contents of the basin and mix with the ash. If the weight of ash is greater than 0.1 g. the amount of flux to be mixed is increased to give a flux to ash ratio of 2:1.

9. Place the basin and contents in a muffle furnace maintained at temperature of 900 to 950°C for five minutes. Remove the basin and swirl the contents to ensure contact of the flux with the ash. Replace the basin in the muffle furnace and maintained at a temperature of 925 +/- 25°C for a further 10 minutes.
10. Cool the fusion melt to room temperature and add 25 ml of the tartaric and hydrochloric acids mixture. Carefully place magnetic follower in the basin.
11. Place basin and contents on magnetic stirrer hot plate. Stir and maintain at 70°C until dissolved.
12. Allow the solution to cool and then transfer it to a 50 ml graduated flask with distilled or deionized water, washing the basin several times to ensure transfer is complete. Make up to the mark with distilled or deionized water. Transfer immediately to the polystyrene test-tube to avoid absorption of silica from the glass.

5.2 - Sulphated Ash. (Method B)

Follow steps 1 to 3 in Section A above, then

4. Add 1 ml of concentrated sulphuric acid for each gram of sample into the platinum basin.
5. Place the basin on a hot plate and increase the temperature steadily, trying to avoid excess foaming while continuously stirring with a PTFE rod. Remove basin from hot plate if excessive foaming develops and return when foaming has subsided. Continually check each sample until foaming has ceased and the acid sludge becomes thicker.
6. Continue to heat on the hotplate until fuming has ceased.
7. Place basin on a porcelain triangle and heat gently trying to ignite any hydrocarbon vapour present until dry coke is obtained. Note. Do not heat crucible even to "dull red" heat, as Vanadium volatilizes and is lost.
8. Place basin in a furnace at a temperature of 150°C and increase slowly to 525°C while introducing a slow flow of air into the furnace to aid ashing.
9. After sample has been completely ashed, remove, cool and reweigh.
10. Place basin on hotplate and add approximately 10 ml of water, 5 ml of concentrated hydrochloric acid and cover with a watchglass.
11. Rinse a 50 ml volumetric flask and stopper three times with distilled water.
12. Filter the acid extract through a Whatman No.1 paper into the 50 ml volumetric flask and rinse the watchglass and basin transferring the washing to the paper.
13. Continue washing the paper until the volume in the flask is approximately 45 mls.
14. Dilute to the mark and mix, mark as acid extract).
15. Transfer the paper to a platinum crucible. Dry, char and ignite (550°C) the paper until all the carbon has been removed.
16. Add 0.2 g Sodium Potassium Carbonate mixture (50/50 Na_2CO_3 / K_2CO_3) and fuse at 750°C for approximately 30 minutes.

17. Extract the cooled melt with distilled water. If the solution contains undissolved material (metal carbonates) transfer it to a 50 ml beaker and dilute to approximately 25 ml with rinsings from the crucible. Acidfy the solution by adding concentrated Hydrochloric acid dropwise while stirring the solution with a PTFE rod. Gently warm the solution to remove the carbon dioxide and, once cool, transfer to a clean 50 ml volumetric flask. Dilute to the mark, mix well and identify with the sample reference number and marking this solution as "Fusion".

5.3. - <u>Ashing and Acid Extraction</u>.(Method C)
Follow steps 1 to 9 in Method A above
and steps 9 to 17 in Method B above.

5.4. - <u>Ashing and Hydro Fluoric Acid Extraction.</u>
According to IP 363 procedure.

5.5. - <u>Precautionary Note</u>
Direct spray of fuel sample after dilution with xylene cannot be used for marine fuels due to the possible presence of cat. fines containing Al and Si which would obstruct the nebuliser. Any solid particles in the sample should be avoided to reduce the possibility of blocking the nebuliser. In case of blockage, clean the nebulizer as described by Meinhard (12).

6. RESULTS AND DISCUSSION

The results obtained by ICPES analysis and varying the extraction method, (including tetraborate fusion) are compared in this section with results on the same samples given by AAS analysis using the solubilising technique appropriate to the element.

6.1.   Comparison of Elemental Analysis by ICPES
        2 samples by 3 extraction methods.

Table 3 gives the results of analysing for the 15 elements Zn, Na, Ni, Ba, K, Fe, Mn, P, Si, Al, V, Ca, Sb, Cu, using ICPES and the following 3 digestion methods.
    A - tetraborate fusion
    B - sulphuric acid ashing
    C - ashing followed by acid extraction
It is noteworthy that 27 values out of 30 obtained by tetraborate fusion and extraction are the highest determined.
    A series of comparisons were arranged between the various procedures on identical samples in two laboratories to demonstrate the feasibility and accuracy of ICPES for simultaneous analysis of trace elements in fuel oils.
    The major exceptions are the phosphorus P result of sample 2 which is attributed to contamination and the nickel Ni figure where the tetraborate result is 4.4 lower than by the sulphated ash result which may have been due to contact with a nickel spatula for example. Otherwise the results for Zn, and Na, differ by only 0.2 and 0.4 ppm respectively.

The overall pattern of comparison between tetraborate fusion and the ash extraction methods indicates that tetraborate fusion effectively solubilises the elements present in ashed material from fuel oil samples.

6.2. Comparison of 2 Digestion Methods. 6 Samples in 2 laboratories.

TABLE 4 gives a comparison of ICP analysis on 6 fuel oil samples in 2 laboratories employing different techniques for solubulising the elements.

The concentration of 11 elements were determined in the 6 samples which in duplicate were subjected to tetraborate fusion and extraction according to IP 363.

The values obtained by the tetraborate fusion are generally higher on the same sample than for extraction by IP 363. This lends more support that the tetraborate fusion technique comprehensively serves to solubilise the analyte elements.

6.3 Comparison of Si, Al and Na by ICP and AAS 5 Samples in 2 laboratories

TABLE 5 compares Si, Al, and Na results on 5 samples using ICP and tetraborate fusion in laboratory I and extraction according to IP 363 with both ICP and AAS in Laboratory II.

All the aluminium results from laboratory I are highest and 4 out of 5 silicon results are highest.

The results from IP 363 extraction by ICP and AAS are in general agreement and would be considered accurate but for the comparison with the results from tetraborate fusion.

The Si and Al results for the 2 extraction methods determined by ICP in the two laboratories differ on average for Si by 4.6 ppm for a range of values from 0.8 to 40.6 & for Al by 5.1 ppm for a range of values from 1.3 to 34.6.

The significant difference in results from the two extraction methods gives further support to the use of tetraborate fusion with 2:1 ratio of tetraborate to ash to completely solubilise the elements.

6.4 Repeatability check of ICPES and Tetraborate Fusion

TABLE 6. Septupilate Determination on 12 elements.

As a check on the repeatibility of the ICPES analysis, 7 samples of the same fuel were analysed for the 12 elements. Si, Al, V, Na, Fe, P, Pb, Ca, Zn, Cu, Sb, Mn. using the tetraborate fusion technique.

The results as tabulated show consistent figures.

Calculation of the average and its standard deviation were made in order to derive a repeatability figure for each element.

The repeatability was defined as
$$\frac{2 * \text{standard deviation}}{\text{average result}}$$

For all 12 elements the repeatability was less than 1 ppm
for 6 elements less than 0.1 ppm and
for 5 elements between 0.1 ppm & 0.5 ppm.
(The repeatability figure for Cu at 0.82 is due to the high result of 0.7 ppm giving an average 0.4 for 7

results. Taking only 6 results the average would be 0.35
and the repeatability 0.3).

6.5   Comparison of ICP analysis with Colorimetric and AAS.

TABLE 7. 2 samples for Si, Al and V by ICP and
Colorometric and for Fe, Na and Ca by ICP and AAS.

Two samples were analysed for Si, Al, V, Fe, Na, and
Ca. The ICP analysis was carried out with the digestion
method A outlined in the paper using lithium tetraborate and
all the results were obtained simultaneously.
For the alternative methods it was necessary to use
four procedures. The samples for colorimetric analysis and
atomic absorption were also digested by the tetraborate
method A and
   The elements Si, Al and V were determined
       colorimetrically by the following methods:
       Silicon using molybdenum blue [13].
       Aluminium measured in sample solution after adjusting
       the pH following colour development with ALUMINON
       (ammonium aurine tricarboxylate) [14].
       Vanadium using ASTM method D 1548 (colour development
       section) [15].
   The elements Fe, Ca and Na were determined by Atomic
       Absorption Spectroscopy.

       Iron and Calcium were determined by flame AAS. using
       a hollow cathode lamp with air/acetylene flame.
       Sodium was determined by flame emission using an
       air/acetylene flame alone.

Note. To eliminate interferences of Na determination by Fe,
       V etc., the solution obtained by tetraborate
       digestion (method A) was analysed in duplicate using
       a standard addition technique involving addition of 1
       ppm Na, to one portion of the solution before AAS
       analysis.

7. APPLICATION OF ICPES ANALYSIS TO MARINE FUELS

      The availability of rapid accurate data on the
presence of metal elements in fuel oils enables diagnosis of
critical levels of both natural constituents and contaminant
levels of harmful elements. As a result advice can be given
to the users of the fuel which is directed to optimising use
of the fuel or minimising the adverse consequences of using
a particular fuel.
      Table 8 gives examples of the elements which may be
found in typical marine fuels with the range of
concentrations encountered in a recent month in a random
batch of 112 samples taken world-wide.
      In these analyses of marine fuels, the following elements
have some significance:
V and Na levels together with the V/Na ratio are important
          for giving advice on the possibility and prevention
          of hot corrosion of exhaust valves and fouling of
          turbo-chargers.

Al is an indicator of the level of catalytic fines in a fuel, as Al is a unique constituent of catalytic cracking catalyst. Levels above 30 ppm in a fuel may give rise to difficulty in purifying the fuel to an acceptable level for the engine, especially if there is any deficiency in the stability of the fuel so that it is prone to form sludge at the temperature of the centrifuge.

Si is also associated with cat fines but is less specific as this element is present in other natural debris such as sand. When present with Al it is confirmatory of the presence of catalytic fines but alone it only gives indication that other debris such as sand may be present.

Pb arises from the presence of waste lub oils which have been drained from gasoline engines. Excessive amounts (above 30 ppm) can result in further modification of ash deposits with aggravation of the hot corrosion effects due to V and Na. Also the nature of the ash can cause some concern if it has to be removed manually during maintenance, particularly in the quantities encountered after a fuel containing lead compounds has been burnt continuously in boilers.

Ca arises from the presence of waste lub oil taken from diesel engines and can improve combustion of a fuel, but adds to the ash content after combustion.

Fe is an indicator of metal contamination by rust,. which can give rise to harmful wear and deposits in combustion spaces

With the availability of rapid accurate data on the concentration of these trace elements present in commercial marine fuels, it is now possible to advise on preventive action in the event that contaminated fuels are delivered to a ship.

The availability of accurate multi-element analysis within 12 hours of receiving a marine fuel sample has made possible the diagnosis of potentially difficult combinations of metals and opens the way to preventive action to minimise contamination and uncontrolled disposal of waste materials. In cases where excessive amounts of particular elements are found, remedial action can be advised promptly to minimise potential hazards from using the fuel.

8. CONCLUSION

Inductively coupled plasma emission spectroscopy (ICPES) is an analytical tool for rapid and accurate determination of up to 25 elements in fuel oil, providing the elements are completely solubilised.

The tetraborate fusion technique with 2:1 ratio tetraborate to ash, provides the requisite solubilisation of all potential elements ensuring no loss of elements or sensitivity. Interference from other elements can be controlled and repeatable results are obtained.

The combination of fuel ashing and tetraborate fusion followed by ICPES analysis provides the rapid, precise, and simultaneous multi-element analysis (with minimum risk of contamination), which is required for effective marine fuel analysis.

ACKNOWLEGEMENT

The authors gratefully acknowlege the support in making available the resources and data necessary to prepare this paper given by
Caleb Brett Laboratories Limited
and
Lloyd's Register of Shipping,
Fuel Oil Bunker Analysis and Advisory Service

TABLE 1

WAVELENGTHS AND DETECTION LIMITS OF ICP QUANTOMETER.

ELEMENT	SYMBOL	WAVELENGTH(nm)	DETECTION LIMIT(mg/kg)
Aluminium	Al	308.2	0.02
Antimony	Sb	206.8	0.05
Arsenic	As	193.7	0.01
Barium	Ba	455.4	0.0003
Beryllium	Be	313.0	0.0002
Bismuth	Bi	223.0	0.0002
Boron	B	249.7	0.004
Cadmium	Cd	226.5	0.002
Calcium	Ca	317.9	0.004
Chromium	Cr	267.7	0.004
Cobalt	Co	228.6	0.004
Copper	Cu	324.7	0.001
Iron	Fe	259.9	0.002
Lead	Pb	220.3	0.02
Lithium	Li	670.7	0.002
Magnesium	Mg	279.0	0.01
Manganese	Mn	257.6	0.0003
Mercury	Hg	184.9	0.007
Molybdenum	Mo	281.6	0.004
Nickel	Ni	231.6	0.008
Phosphorus	P	178.2	0.05
Potassium	K	766.4	0.2
Silicon	Si	288.1	0.006
Sodium	Na	589.0	0.1
Strontium	Sr	407.7	0.00028
Sulphur	S	180.7	0.08
Tin	Sn	189.9	0.04
Titanium	Ti	337.2	0.001
Tungsten	W	239.7	0.02
Vanadium	V	311.0	0.003
Zinc	Zn	202.5	0.001

TABLE 2

DESCRIPTION of ICPES INSTRUMENTATION & OPERATING CONDITIONS

Determination		Simultaneous
Forward power	kW	1.30
Frequency	MHz	27.12
Viewing height above load coil	mm	14
Viewing window, square of side	mm	4
Torch type		Fassel
Gas flow rates	l/min	
Coolant		12
Auxiliary		0.8
Injector (humidified)		1.0
Spray chamber		Conical single pass
Nebuliser type		Concentric glass Meinhard TR-30-A3
Solution uptake rate (unpumped)	ml/min	2.5
Uptake tube (polythene) length * i.d.	mm	350 * 0.5
Preflush time	secs	20
Detection sequence number * time of integration	secs	3 * 5

TABLE 3

ELEMENTAL ANALYSIS of 2 FUEL OIL SAMPLES by ICPES.

Comparison of 3 extraction methods (Rounded to 1 decimal)
- A. metaborate fusion
- B. sulphuric acid ashing
- C. ashing followed by acid extraction.

SAMPLE 1.

Extractn Method	Zn	Na	Pb	Ni	Ba	K	Fe
A	0.3	71.8	1.4	12.8	0.6	<0.4	22.0
B	0.2	71.1	1.1	6.0	0.6	1.3	15.7
C	0.4	65.2	0.9	6.1	0.5	0.7	15.6

	Mn	P	Si	Al	V	Ca	Sb	Cu
A	0.2	0.9	64.0	61.9	20.2	5.8	9.4	0.4
B	0.2	0.7	54.9	44.8	19.0	3.7	10.5	0.2
C	0.1	0.2	56.1	45.2	17.2	5.1	7.6	0.1

SAMPLE 2.

Extrctn Method	Zn	Na	Pb	Ni	Ba	K	Fe
A	3.5	37.7	1.3	17.0	0.8	1.5	38.2
B	3.7	38.1	1.0	21.4	0.8	1.3	34.3
C	3.6	36.9	0.9	13.0	0.7	3.1	32.1

	Mn	P	Si	Al	V	Ca	Sb	Cu
A	0.4	9.5	20.4	10.9	41.0	19.8	0.23	0.5
B	0.3	22.8	6.1	7.8	39.0	17.3	<1	0.5
C	0.3	7.9	9.9	6.9	35.5	17.1	<1	0.4

Note All values for metaborate extraction are in general the highest determined, indicating analysis was complete. The Potassium results are the exception due to chemical contamination by K during acid extraction

TABLE 4

COMPARISON of ICP ANALYSIS for 11 elements

6 fuel oil samples using 2 digestion methods
in different laboratories

Lab I   Tetraborate Fusion according to Procedure A.

Lab II   Extraction according to IP 363

SAMPLE	ELEMENT Lab	Si	Al	V	Fe	P	Pb
812	I	0.8	1.3	41.6	4.7	0.5	0.1
	II	2.8	0.8	42.7	4.6	0.4	0.6
816	I	5.7	6.1	100.3	18.1	1.4	0.3
	II	4.9	2.8	100.1	15.9	0.7	1.0
819	I	29.8	19.1	28.0	38.1	3.1	0.3
	II	22.4	14.1	31.2	35.8	1.9	0.4
820	I	16.6	14.9	42.1	8.9	1.2	0.4
	II	11.1	9.4	43.9	6.6	0.3	0.6
822	I	1.3	1.0	35.0	5.3	1.0	<0.1
	II	1.2	4.6	35.3	5.1	0.1	0.3
866	I	40.6	34.6	37.2	56.3	44.8	20.1
	II	29.1	21.2	34.5	45.7	24.9	9.5

SAMPLE	ELEMENT Lab	Ca	Zn	Cu	Sb	Mn
812	I	2.2	1.2	1.2	0.1	0.1
	II	2.1	1.1	0.5	0.1	<0.1
816	I	2.3	1.0	0.7	0.1	0.1
	II	2.0	1.1	0.4	0.4	0.1
819	I	4.2	2.0	0.4	0.2	0.3
	II	4.0	2.1	0.3	<0.1	0.3
820	I	1.7	0.5	0.7	0.4	0.1
	II	1.0	0.5	0.3	0.1	<0.1
822	I	1.5	0.6	0.2	<0.1	<0.1
	II	0.9	0.5	0.2	0.1	<0.1
866	I	39.0	33.3	3.2	0.3	0.6
	II	9.3	29.6	2.0	<0.1	0.5

TABLE 5

ANALYSIS of SILICON, ALUMINIUM and SODIUM

COMPARISON of 5 SAMPLES by ICP and AAS.

LAB	INSTRUMENT	Si	Al	Na	Si	Al	Na
	SAMPLE		812			816	
I	ARL ICP	0.8	1.3	24.3	5.7	6.1	13.3
II	PE ICP	2.8	0.8		4.9	2.8	
II	PE AAS	5.3	0.8	26.6	6.6	3.1	14.1
	SAMPLE		819			820	
I	ARL ICP	29.8	19.1	15.6	16.6	14.9	14.8
II	PE ICP	22.4	14.1		11.1	9.4	
II	PE AAS	18.3	13.8	15.9	11.3	10.1	14.4
	SAMPLE		866				
I	ARL ICP	40.6	34.6	39.4			
II	PE ICP	29.1	23.5				
II	PE AAS	28.9	23.1	30.8			

Laboratory I Lithium Tetraborate fusion, extraction by tartaric acid, hydrochloric acid mixture.

Laboratory II Hydrofluoric acid extraction according to method No IP 363

ARL = Applied Research Laboratories   PE = Perkin Elmer

TABLE 6

COMPARISON of 7 REPLICATE DETERMINATIONS on 12 ELEMENTS

7 Samples of the same fuel ashed and fused with
Lithium tetraborate for analysis by ICPES.

Sample Weight	Si	Al	V	Na	Fe	P
16.22	21.3	18.9	217.3	63.3	15.8	1.0
15.37	21.7	19.2	216.6	61.6	15.9	1.1
16.41	21.7	19.2	216.4	61.9	17.8	1.0
15.47	22.5	19.6	221.6	63.0	24.6	1.2
19.36	21.7	18.8	212.2	60.8	16.7	1.2
15.36	21.5	19.3	208.2	58.0	18.0	1.2
15.40	21.8	19.2	217.8	60.6	16.8	1.3
Average	21.74	19.17	215.7	61.31	17.9	1.14
Std Dev	0.37	0.26	4.31	1.78	3.05	0.11
Rptblty	0.03	0.03	0.04	0.06	0.34	0.20

Sample Weight	Pb	Ca	Zn	Cu	Sb	Mn
16.22	0.5	6.1	0.7	0.2	0.9	0.2
15.37	0.7	6.5	0.7	0.3	1.0	0.2
16.41	0.6	6.3	0.7	0.3	0.9	0.2
15.47	0.7	6.6	0.8	0.4	1.0	0.2
19.36	0.6	6.4	0.8	0.5	0.9	0.2
15.36	0.9	6.4	0.8	0.7	1.2	0.2
15.40	0.9	6.5	0.8	0.4	1.1	0.2
Average	0.7	6.4	0.76	0.4	1.0	0.2
Std Dev	0.15	0.16	0.05	0.16	0.12	0.0
Rptblty	0.44	0.05	0.14	0.82	0.23	0.0

Repeatability = $\dfrac{2 * \text{Standard Deviation}}{\text{Average result}}$

TABLE 7

COMPARISON OF ICP ANALYSIS WITH COLOROMETRIC AND AAS RESULTS

Two samples were analysed for Si, Al, V, Fe, Na, Ca. in duplicate using ICPES and compared with Colorimetric & AAS.

		Si	Al	V
Sample I	ICP	2.7	3.9	95.2
	Colorometric	2.4	4.0	94.8
Sample II	ICP	26.1	16.1	122.8
	Colorometric	25.3	15.8	121.3

		Fe	Na	Ca
Sample I	ICP	14.4	9.2	1.2
	AAS	14.1	12.7	1.0
Sample II	ICP	14.3	31.6	7.2
	AAS	14.0	33.0	7.5

Note The ICP analysis was carried out with the digestion method A outlined in the paper using lithium tetraborate and all the results were obtained simultaneously.
For the alternative methods it was necessary to use the following four procedures:-

COLOROMETRIC METHODS.

Silicon.	Molybdenum blue method	Reference (13)
Aluminium.	ALUMINON method	Reference (14)
Vanadium	ASTM D-1548 Color development Section Ref (15)

AAS METHODS

Iron &)
Calcium)   Flame AAS with hollow cathode lamp
Sodium   Flame emission by standard addition

TABLE 8.

Elemental Analysis of Marine Fuels

High and Low Values of 12 elements taken from
112 typical fuel analyses world-wide.

Element		Low value	High value
Silicon	Si	1	27
Aluminium	Al	0.5	22
Vanadium	V	0.2	219
Sodium	Na	0.9	86
Iron	Fe	3	382
Phosphorus	P	0.6	277
Lead	Pb	<0.1	35
Calcium	Ca	7	224
Zinc	Zn	0.4	32
Copper	Cu	0.4	36
Antimony	Sb	<0.1	0.4
Manganese	Mn	0.1	38

REFERENCES

1. Allemand, C.D. ICP Information Newsletter 1976,2, 1-26
   Design of an ICP discharge system for direct residual fuel analysis for marine applications.

2. Fassel, V.A., Peterson C.A. and Abercrombie F.W.
   Anal. Chem. 1976, 48, 516-519.
   Simultaneous determination of wear metals in lubricating oils by ICPES.

3. Saba, C.S., Rhine, W.E. and Eisentraut, K.J.
   Anal. Chem. 1981, 53, 1099-1103.

4. Boorn, A.W. and Browner, R.F.
   Anal. Chem. 1982, 54, 1402-1410.

5. Dolezal, J., Povondra, P. and Sulcek, Z.
   Decomposition Techniques in inorganic analysis,1968,P.91
   Iliffe Books Ltd. London.

6. Burman, J.O., Ponter C. and Bostrom K.
   Anal. Chem. 1978, 50, 679.
   Metaborate digestion procedure for ICP. optical emission spectrometry.

7. Greenfield, S., Proc. Analyt. Div. Chem. Soc. 1976, Page.279-284. Why Plasma Torches? - Invited Lecture.

8. Greenfield, S., Jones, I.L.,McGeachin, H., MCD, and Smith, P.B. Anal. Chim. Acta, 1975, 74, 225-245.
   Automatic, Multi-Sample, Simultaneous Multi-element Analysis with a H.F. Plasma Torch and direct reading Spectrometer.

9. Wendt, R.H., and Fassel, V.A.
   Analyt.Chem. 1965, 37, 920-922.
   Induction-Coupled Plasma Spectrometric Excitation Source.

10. West, D.C., and Hume, D.N., Analyt.Chem.1964, 36,412-415
    Radio Frequency Plasma Emmission Spectrophotometer.

11. Boumans, P.W.J.M. and Deboer, F.J.,
    Spectrochim. Acta, 1975, 308, 309 -334.
    Studies of an inductively-coupled high-frequency Argon Plasma for optica/emission spectrometry II - Compromise conditions for simultaneous multi-element analysis.

12. Meinhard,J.E. ICP Information Newsletter 1987,12,677-680.

13. Analysis of Raw, potable and waste water 1972, Page 71.
    Her Majesty's Stationery Office, London.

14. Ibid.Page 69.

15. Annual Book of ASTM Standards, 1987. Method D-1548.

HPLC Determination of Aromatics in Middle/Heavy Distillates
A Computerized Approach

P. Richards
Carless Solvents Ltd, Harwich Refinery, Refinery Road, Parkeston, Harwich, Essex, CO12 4SS

It has been, for many years, our policy to train our Operators to carry out their own Quality Control, by the use of ASTM/IP test procedures where appropriate, and in house GLC, IR/UV methods when not. Quality Assurance is carried out by the Laboratory Staff on all final products.

The determination of aromatic contents of middle and heavy distillates has always been difficult. The problems increase with boiling point due to increased complexity of the matrix, which leads to a divergence in results obtained by differing methods.

The test requirements of a new dearomatization plant dictated that a suitable method for the measurement of aromatics in middle and heavy distillates be found.

We first had to define the requirements of the method to be used. These were:-

1. Safe for Process Operators to carry out.

2. Simple for Process Operators to be trained to use.

3. Easy to calibrate.

4. Cover the boiling range of 130–400°C.

5. Quick i.e. less than 30 minutes.

6. Non labour intensive.

The methods available at the time were:-

1. I.P. 145   Sulphonation
2. I.P. 156   Fluorescent Indicator Adsorption.
3. Infrared   Brandes.
4. Ultraviolet   In House.
5. G.L.C.

1. SULPHONATION

This simple but very reliable method involved reacting the sample under test with a mixture of Sulphuric Acid and Phosphorous Pentoxide at 0°C, separation being achieved by centrifuging, the residual hydrocarbon volume measured, and the aromatic content then calculated.

This method we deemed as unsuitable for our needs on safety grounds. I.P./ASTM have since come to the same conclusion and deleted the method.

2. FLUORESCENT INDICATOR ADSORPTION

A displacement chromatography method of dubious accuracy, which involves preparing a silica gel column to which has been added a small amount of fluorescent markers, eluting the sample with isopropanol and physically measuring the band lengths. The result is normalised against the total sample length. This method fell short of our requirements on three counts:-

    (a)   Labour intensive (column packing)
    (b)   Inaccurate   (measuring bands)
    (c)   Inherently poor separation of aromatics.

It takes considerable time to activate the silica gel and it requires mixing during the heating stage to ensure that a consistent batch is achieved. The packing of a column, although simple for skilled laboratory staff, does not lend itself to being carried out by Process Operators.

The measurement of the bands achieved is purely manual and if tailing occurs, which is often the case, a value judgement has to be made. This is unacceptable for an "On Stream" test.

Analytical work with the use of internal standards followed by extraction and analysis on a Gas Liquid Chromatograph, showed the separation between the aromatic and paraffins bands was incomplete. The aromatic band only contained between 75 and 96% actual aromatic compounds, thus over-estimating the aromatic content. The aromatic band was also found to have an extremely variable concentration on duplicate results.

3. INFRARED - BRANDES

The Infrared absorption is measured at $1610 cm^{-1}$ and the results calculated using a given formula. The formula includes a constant of 1.2 and is therefore unsuitable for samples with less than about 15% aromatics, also its use on middle distillates is outside the scope, as it was originally calibrated for lube oil.

Calibration against FIA gave reasonable results but matrix changes made constant recalibration necessary. In specific applications it is only as accurate as the FIA result, which as mentioned earlier is dubious.

4. ULTRAVIOLET

The vast differences in the extinction coefficients between aromatic compounds makes the use of this method unsuitable for materials with more than 0.1% aromatics.

5. G.L.C.

Gas liquid chromatography is an ideal method for light distillates where individual components can be separated identified and quantified, but the complexity of the matrix increases in the middle and heavy distillate ranges and individual identification becomes impossible unless a mass spectrograph is employed.

This last highly expensive and sophisticated technique did not appear as a viable solution for Process Operators, even though they regularly use capilliary GLC analyses on light distillates.

CONCLUSION

As none of these methods fulfilled the requirements we needed to a new method was needed. Preliminary investigations with the aid of Perkin-Elmer indicated that HPLC had the potential to meet our requirements.

6. HPLC

The two most commonly used detectors for aromatics are Ultraviolet and Refractive Index.

(a) The Ultraviolet detector is extremely sensitive and selective, but is really only suitable for measuring known aromatics or standardised matrices. The current tentative IP method uses a U/V detector to measure only the bicyclic aromatics. Mixtures are difficult due to the vast differences in extinction coefficients as mentioned earlier.

(b) The Refractive Index Detector is less sensitive but the mass dependant properties of materials ensure the response for mixtures of aromatics are far more uniform than with Ultraviolet. It is, however, very sensitive to temperature, pressure and flow variations.

The column suggested by Perkin-Elmer was Spherisorb 5 Amino 25cm long. This column was selected due to its ability to resolve paraffins and aromatics completely. Work carried out by Patrick Grizzle and Donna Sablotny has shown that not only does this column separate aromatics from paraffins completely but separates the aromatics into ring order. This work has now been published in Analytical Chemistry[1].

Our original work with this column system indicated that the chromatography needed to be improved. With synthetic mixtures the paraffins and monocyclic aromatics were resolved. This was not the case with real samples. To obtain as near as possible base line separation it was necessary to increase the column length and decrease the porosity. A Spherisorb 3 Amino column, 50 cms length was, therefore, used which gave an improved resolution.

The solvent delivery system was a 25 litre stainless steel can connected to a 1 micron hydrophobic filter and pump. The solvent employed was 65/70 hexane (Food Grade) helium degassed and delivered to the pump via double-pass debubblers under a slight pressure of helium.

The standard means of filling the reference side of detectors has been manual by syringe or pump followed by connecting the inlet and outlets together. This was soon found to be inadequate for extended duration and we had to resort to a delivery system similar to the solvent analytical side but at very low flow rates, with a reservoir of degassed hexane connected via a debubbler to the reference side of the detector. The detector and column were thermostatically controlled at 25°C with a circulating water system. The sample to be tested was introduced via a loop injection valve and the analogue signal generated by the detector was collected by an Analogue/Digital interface and the data manipulated by a computer.

With this configuration a stable base line can be achieved running continuously for several weeks. We have increased the solvent delivery system to 50 litres since initial commissioning, thus making calibration checks and refilling only necessary at 6 weekly intervals.

CALIBRATION

Initial commissioning required that the equipment should be on line virtually as soon as it arrived. Taking a simplistic approach and plotting integrated areas against aromatic content gave a plot that was not linear enough to use a simple factor to calculate the aromatic content. Using an Nth order regression gave a polynomial of the fourth order with a good fit from 1-50%. Below 1% the accuracy was questionable due to the inclusion of a small constant. The situation was compounded by the fact that the eluent being used was of inconsistent quality. The naphthene content varied thus the density and refractive index changed requiring a full calibration and polynominal calculation. We needed a more general curve fit to reduce the calibration time.

It was more difficult to obtain a computer solution to this than for apparently more complex matrices such as ASTM D2140 and D2159, both of which had already solved. The published work has used mainly recorders and calculations based on peak height measurements.

A relationship was required between density and refractive index and Newton's formula was employed viz:-

$$\frac{d}{R^2 - 1} = \text{constant}$$

where d = material density
R = refractive index

A later empirical formula by Ekyman has also been investigated but apparently gives little advantage over the Newtonian solution[2].

The incorporation of this relationship into a useable calculation required three basic assumptions.

1. The following conditions used were assumed to be constant; pressure, flow and injection volume. The injection volume should also be that of the cell and if not then a correction has to be applied.

2. No volume change occurred.

3. The aromatic concentration in the sample was transmutable to a concentration % v/v in the detector.

Using the formula

$$\frac{100}{\text{Density}_T} = \frac{\text{Aromatic \% wt}}{\text{Density}_A} + \frac{100 - \text{Aromatic}}{\text{Density}_S}$$

and substituting the Newtonian equation an expression can be derived to connect concentration and differential refractive index.

This expression was then manipulated to give a relationship between the signal output and the concentrations. The intention was to integrate this formula over the whole of the peak against time, which gave :-

$$\int_{t_1}^{t_2} V\, dt = \int_{t_1}^{t_2} \left[\sqrt{\left(\frac{100\,(R_a^2-1)\,(R_s^2-1)}{A_c(R_2^2-1)+(100-A_c)(R_A^2-1)} + 1\right)} - R_s \right] dt$$

where $\int_{t_1}^{t_2} V\, dt$ = Integrated area

This cannot be integrated as there is no formula to connect aromatic concentration with time. It was suggested one route would be to use the solution of a Gaussian curve and insert it into the formula, another was to obtain a set of constants by calibrations[3].

We chose the latter approach and plotting the constants obtained from equation (4) against area found. A straight line was obtained (3), which by a simple step incorporates into the calculation (1), and converts back to percentage v/v (2).

1. $$A_w = \left(\frac{100\,(R^2_{ac} - 1)}{R^2_{ac} - R^2_s}\right)\left(1 - \left(\frac{R^2_s - 1}{(R_s + f_k I)^2 - 1}\right)\right)$$

2. $$A_v = \frac{100\,D_S\,A_w}{100\,D_{AC} - A_w\,(D_{ac} - D_s)}$$

3. $f_k = XI_c + Y$

   where   X + Y are the contants of calibrations

4. $$K = \left(\frac{1}{I_c}\right)\left(\left(\sqrt{\frac{100\,(R^2_a - 1)(R_2 s - 1)}{A_{cw}(R^2_s - R^2_a) + 100\,(R^2_a - 1)}} + 1\right) - R_s\right)$$

   where

5. $$A_{cw} = \frac{100\,A_{cw} \times D_a}{A_{cv}D_A + (100 - A_{cv})\,D_S}$$

I_C = Integrated curve area.
A_{cw} = Aromatic content of standard % w/w in cell.
A_{cv} = Aromatic content of standard % v/v
R_A = Refractive index of aromatic standard
R_S = Refractive index of solvent
D_a = Density of the aromatic standard
D_S = Density of the solvent

D_{ac} is the density of the calibration ring type aromatics
R_{ac} is the refractive index of the calibration ring type aromatics.

I = Integrated area
A_w = % w/w aromatic in cell
A_v = % v/v in sample

The fit obtained was much better between 0.05 and 1% than the polynomial and almost identical for the range 1-50%. The calculation is in fact still using a fourth order polynomial but does not involve a simple addition with a constant.

Once the linearity of the detector has been established calibration can be carried out at two points with an intermediate check. This is not the case with a simple polynomial which requires multiple calibration points.

COMPUTER PROGRAMME

The type of computer being used to obtain the integrated areas had a BASIC programming facility and was capable of abstracting the data and then carrying out mathematical procedures.

We had for many years used several extensive data manipulation routines on GLC data to obtain calculated figures for Motor/Research Octane Ratings, RVP, density and molecular weight. The requirement to write a programme that could be run by Process Operators automatically without reference to Chemists meant that we had to incorporate traps to filter out spurious data.

The programme relies upon the original integrated areas and retention times and by the use of flags calculation of the relative retention times.

The flag is originally set to a value of -1 and is reset to the retention time of the first aromatic peak by setting a high minimum area for the initial paraffin peak, which is normally very large thus excluding any spurious peaks or noise prior to the analysis. Once this is found the flag is set to zero. The next peak is the first aromatic peak when the flag is set to its retention time.

Each peak is considered separately. The flag is used as a reference time and the type of aromatic is calculated by using Boolean Logical Operators. This means that in one line several decisions can be made.

Once the type of aromatic is known the correct calibration is used and the result added to an array.

Changing from the original polynomial to the more complex formula presented no computing problems as this only meant including a few extra lines.

We have transferred the programme from a HP3353E Lab data system to a Trivector Trio since the original work. This was relatively simple as the main difference between the BASICS (Hewlett-Packard and BBC) is in string handling.

The chromatographic separation required to resolve paraffins from monocyclic aromatics also resolves the bicyclic and tricyclic aromatics as mentioned earlier. We can within 30 minutes give a good breakdown of the aromatic types present.

Results have been correlated to GLC mass spectroscopy results on three separate occasions, through our own work, a Round Robin by the API, and MAFF. It is interesting to note at this stage that the GLC did not find the tricyclic components in the API work as they were not expected to be present and the conditions employed were not exacting enough. This has since been rectified.

Modification to our standard technique has allowed us to carry out determination of polycyclic aromatics and correlation to an in line Ultraviolet detector.

The terms mono, bi, tri and polycyclic aromatics previously referred to are for fused benzene ring structures. Compounds which differ, for instance those containing five membered fused rings or bridges can elute, as shown by Grizzle and Sablotny, in apparently anomalous bands e.g. Fluorene, a tricyclic compound containing a five membered ring and two benzene rings, elutes at the end of the bicyclic band and is so allocated. Biphenyl on the other hand elutes within the monocyclic aromatic band.

Improvements in integrators available means that we now have a dedicated computer for both detectors. The one currently employed is the Trivector Trio, which gives an added advantage over the 3353E system in that integration parameters can be changed post analysis and the data re-analysed. This is required when working at the low end concentrations as base line effects can exaggerate the result.

We are sure that this type of method is far superior to the methods mentioned earlier and gives us a means of further mathematical manipulation of the data. We have carried out an exercise to relate the aromatic content to the smoke point of kerosines using a multiple linear regression technique. Using the formula produced we can estimate smoke points reasonably accurately and use the calculated figure for manufacturing Premium Cl Kerosine. This investigation has confirmed the view that the response effect of bicyclic aromatics on the smoke point is considerably greater than that of monocyclic aromatics. The presence of even small amounts of bicyclic aromatics significantly lowers the smoke point.

HPLC Determination of Aromatics in Middle/Heavy Distillates

ACKNOWLEDGMENTS:

    Perkin Elmer   (HPLC Technical Service)

    Mr. A. Ellis   (University of Essex) (Personal contact)

References:-

1. Analytical Chemistry 1986, 58, 2389-2396 (P.Grizzle & Sablotny).
2. "Organic Solvents Physical Propertus and Methods of Purification"(Techniques of Chemistry Volume II), by Riddick, Bunger and Sakano, Wiley Interscience Wiley, 1986.
3. Dr. J. Greenman   (University of Essex) (Communication)

```
/>LIST,999
   10 DIM C(5,5),RSLT$(10),RSLT(5,11)
   20 J$="CYCLIC AROMATIC CONTENT"
   30 *SWAP
   35 PROC_tinit
  100 C(1,3)=1.5055
  110 C(1,4)=.88018
  120 C(2,3)=1.6082
  130 C(2,4)=1.0198
  140 C(3,3)=1.59427
  150 C(3,4)=.9784
  160 FORI=1TO3:C(I,3)=C(I,3)^2:NEXT
  165 FORI=1TO3:C(I,5)=0:NEXT
  170 S=1.3816
  180 S1=.6706
  190 T1=-1
  200 PROC_stat
  210 IF LEFT$(ID$,1)="B"THEN30
  215 *OPT4
  216 @%=&20206
  220 FORI=1TOR
  230    PROC_peak
  235    IF R1<8.1 THEN 380
  240    IFT1=0ANDR2>10000THEN LET T1=R1
  250    IFT1<0ANDR2>300000THENLETT1=0
  260    IFT1<1THEN380
  270    J=1-(R1/T1>1.17)-(R1/T1>1.86)
  275    ON J GOSUB 1000,2000,3000
  280    K=0
  290    FORQ=1TO2
  300       K=K+C(J,Q)*R2^(Q-1)
  310    NEXT
  320    PRINTR1;"      ";K*1E10;"      ";J
  330    R2=R2*K
  340    R2=(1-((S^2-1)/((S+R2)^2-1)))
  350    R2=100*R2*(C(J,3)-1)/(C(J,3)-S^2)
  360    R2=100*R2*S1/(100*C(J,4)-R2*(C(J,4)-S1))
  370    C(J,5)=C(J,5)-R2*(R2>0)
```

```
380 NEXT
400 PRINT"METHOD :- ";M$;TAB(40);"TIME   ";Hr$;":"Min$;"   ";
405 PRINT"Dated    ";Day$;"/";Mon$;"/19";Yr$
410 REM
420 REM
430 PRINT
440 PRINT
450 PRINTTAB(10);"AROMATICS BY LIQUID CHROMATOGRAPHY"
460 PRINTTAB(10);"================================="
470 REM
480 REM
490 PRINT
500 PRINT
510 @%=&20208
515 Z=0
520 FORI=1TOJ
530    Z=Z+C(I,5)
540    ON I GOSUB800,850,900
550    PRINTJ$;
560    Z$=STR$(INT(C(J,5)))
570    PRINTTAB(40-LEN(Z$));C(I,5)
580    PRINT
590 NEXT
600 PRINT"PARAFFIN CONTENT ";
605 Z=100-Z
610 Z$=STR$(INT(Z))
620 PRINTTAB(40-LEN(Z$));Z
630 FORI=1TO9:PRINT:NEXT
700 PROC_scrap
710 @%=0
720 GOTO30
800 PRINT"MONO ";:RETURN
850 PRINT"  B1  ";:RETURN
900 PRINT"  TRI ";:RETURN
```

Ion-Chromatography in the Oil Industry

D. Mealor
B.P. Research Centre, Chertsey Road, Sunbury-on-Thames, Middlesex TW16 7LN

A. INTRODUCTION

Ion chromatography determines inorganic and short carbon chain organic anions and cations in aqueous solution at ppm and lower levels. Figure 1 shows a chromatogram of some of the anions obtained when petrol is combusted in an engine. These ions can cause corrosion and they can be detrimental to lubricating oils. The origin of these ions is fairly obvious: most petrol contains some sulphur, hence the sulphate; the organic acids are from incomplete combustion of the hydrocarbon, organic bromine compounds are added to petrol to help remove the lead from the cylinders, chloride is probably adventitious and the largest concentrations, nitrite and nitrate are from nitrogen compounds in the petrol and from atmospheric nitrogen.

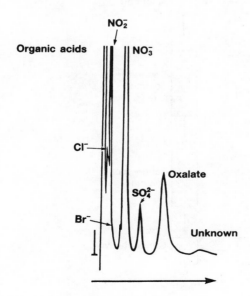

Figure 1. Anions from combustion of petrol.

This example illustrates the utility of ion chromatography in one area of the oil industry, other examples will be given later in the paper. It will continue with a discussion of the techniques and apparatus used for ion chromatography and will conclude with a short discussion of the direction developments in the technique may take.

B. TECHNIQUES AND APPARATUS

1. Suppressed Conductivity Detection

Small, Bauman and Stevens (1) gave the name ion chromatography to the system they published shown in Figure 2. Anions were separated on an ion exchange separator column using a dilute solution of sodium carbonate and bicarbonate as mobile phase. The eluent from this column then passed through a suppressor column which was a high capacity cation exchange resin in the hydrogen form and which removed the sodium to give a solution of carbon dioxide in water. This had only a small conductivity against which the anions, then present as acids, could be detected with great sensitivity by electrical conductivity. Performance data for such a system is given in Table 1.

This system has the great advantage that most ions can be detected by conductivity and so it is a multi-ion technique. An analagous system can be used for the determination of cations such as the alkali and alkali earth metals, ammonia and short chain amines. The mobile phase would then be a dilute solution of a mineral acid.

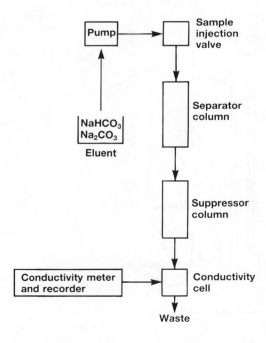

Figure 2. Ion-chromatography flow scheme for Anion analysis.

Ion-Chromatography in the Oil Industry

The mobile phase was very dilute to minimise background conductivity and to avoid frequent regenerations of the suppressor. To enable dilute eluents to be used the separator ion exchange column had to have a correspondingly low capacity. The production of these resins is a complex process, thus the resin for the separation of cations is usually styrene/divinylbenzene copolymer beads which have been sulphonated only on the surface. To make the anion column a latex of anion exchange material is agglomerated onto the surface of the cation exchange resin. It neutralises the sulphonate sites on the surface and provides the layer of anion exchange sites, usually of quaternary ammonium groups. These resins have a capacity of only 0.02-0.05 m equiv/g whereas a fully functionalised resin would have a capacity of about 1 m equiv/g.

The original suppressor column was succeeded by a hollow tube of ion exchange material and then a more robust flat membrane device (2). As illustrated in Figure 3 the unwanted counter ion passes from the mobile phase through the membrane and into a counter flow of regenerant.

TABLE 1

PERFORMANCE DATA

Ion	level mg/l	number of determinations	σ mg/l	detection limit mg/l
Cl^-	3	10	0.020	0.03
NO_2^-	5	"	0.044	0.03
PO_4^{3-}	10	"	0.111	0.04
Br^-	10	"	0.094	0.02
NO_3^-	10	"	0.072	0.03
SO_4^-	10	"	0.053	0.03

Chromatographic conditions:

Guard column	Dionex AG3
Separator column	Dionex AG3
Fibre suppressor	Dionex AF5
Sample loop	100 µl
Flowrate	3 ml.min^{-1}
Mobile phase	2.8 mM $NaHCO_3$, 2.2 mM Na_2CO_3
Conductivity detector range	10 µS

2. Non-Suppressed Conductivity Detection

The suppressed system is proprietary to one manufacturer, others have developed non-suppressed ion chromatography (3). The apparatus is the same as in Figure 2 but without the suppressor. This of course means that the background conductivity is higher. To counteract this there have been some improvements in conductivity detectors to reduce noise and mobile phase concentrations and ion exchange capacities have been further decreased. Although not as sensitive as the suppressed system non-suppressed ion chromatography is useful for many samples, as illustrated by the chromatogram of common anions given in Figure 4, and has some advantages which are given in Table 2.

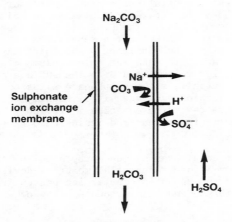

Figure 3. Membrane suppressor.
Eluent from the column flows between the membranes,
regenerant flows in the opposite direction on the outside.

3. Inverse UV Detection

The inverse UV detection method has also been introduced which has equivalent systems for cations and anions (4). For the detection of anions the mobile phase for the ion exchange separation must be a salt of a UV absorbing anion such as potassium phthalate. During the chromatographic run (Figure 5) the concentrations of cations and anions in the mobile phase must be equal to each other and since the potassium concentration is constant the anion concentration must be constant. Thus when the separated anions pass through the detector there is an equivalent decrease in phthalate ion and hence in the UV absorbance. A typical chromatogram is given in Figure 6. The three modes of detection are compared in Table 2.

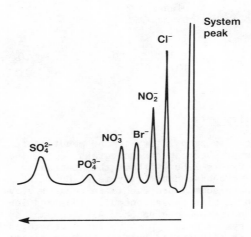

Figure 4. Non-suppressed ion-chromatography.
Waters ion-chromatograph and anion column.

TABLE 2

COMPARISON OF DETECTORS FOR ION CHROMATOGRAPHY

SUPPRESSED CONDUCTIVITY	NON-SUPPRESSED CONDUCTIVITY	INVERSE UV
Low background, high sensitivity	Higher background lower sensitivity but sufficient for many applications	High background lower sensitivity but sufficient for many applic.
Low capacity of analytical column can be oveloaded	Very low capacity, easily overloaded	Very low capacity easily overloaded
Requires suppressor and conductivity detector	Requires conductivity detector	Can be done on standard HPLC with UV detector
Cannot detect anions of very weak acids eg S^{2-} CN^-	Detects all ions	Detects all ions
'Water' dip removed by making samples with eluent	Large system peak	Large system peak
Sensitivities vary	Sensitivities vary	Sensitivity in area/equivalent are constant for all ions
Good possibility for performing gradient elution	No possibility for gradient elution	No possibility for gradient elution

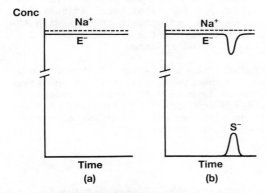

Figure 5. Inverse UV detection.
(a) No sample injected, concentration of Na^+ and eluent anion E^- are constant and equal.
(b) Chromatogram of sample anion S^-, Na^+ remains constant but E^- is decreased when S^- elutes so that the total anion concentration is constant and equal to Na^+.

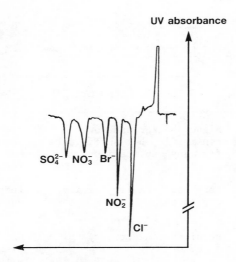

Figure 6. Inverse UV detection.
Vydac silica ion-chromatography column 40 mg/litre of each anion.

4. Other Separation Systems

There are two further systems which are usually classed as ion chromatography.

(a) Ion Pair Chromatography

Ion pair chromatography (5) (often called mobile phase ion chromatography or dynamic ion chromatography) can be performed with all three of the detection methods given above. It is carried out under so called reverse phase conditions and uses a non polar stationary phase such as an unfunctionalised styrene/divinyl benzene resin or a C_{18} silica based material. The principle of the method is that the mobile phase contains an ion pairing agent, such as butyl ammonium hydroxide, which forms a neutral hydrophobic species with the ions to be separated. These partition between the mobile phase and the non-polar stationary phase to different extents.

It is particularly useful for the separation of larger ions, (Figure 7), but can also be used for the common inorganic anions given in Figure 4. Usually the resin phase is used with an alkyl ammonium hydroxide mobile phase and suppressed conductivity detection. The C_{18} silica based column is used with an alkyl ammonium salt and either the inverse UV or non-suppressed conductivity detection.

(b) Ion Exclusion Chromatography

Ion exclusion/affinity chromatography is used for the separation of the anions of weak acids (6). The stationary phase here is a hydrogen form sulphonate type cation exchange resin of high capacity. The mobile phase is usually dilute acid such as 20 ppm HCl. Only neutral species can enter and be retained by the resin. Thus the less ionised the anions (the weaker the equivalent acids) the more

Figure 7. Ion pair chromatography with suppressed conductivity detection.
Dionex MPIC column.
Eluent 30% acetonitrile, 2mM tetra butyl ammonium hydroxide, 1 mM sodium carbonate.

they are retained. A typical separation of the monocarboxylic acids is given in Figure 8. In the example given at the start of this paper ion exclusion chromatography was used to show that the combustion products of petrol only contained the first three of these acids. Weak inorganic acid anions such as fluoride can be determined this way, also borate (Figure 9) if fructose or another 1,2 dihydroxy compound is added to the mobile phase (7) to complex this ion. As the strong mineral acids are eluted together at the solvent front, ion exclusion can be used in combination with ion chromatography to separate both sets of ions (8).

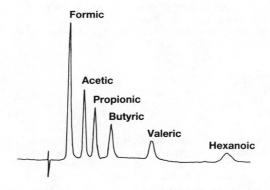

Figure 8. Ion exclusion chromatography.
Aminex A7 column, eluent 20 mg/litre HCl.

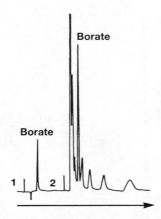

Figure 9. Ion exclusion chromatography of borate.
Aminex A7 column, eluent 0.1 M fructose
1. Standard 10 mg B/L
2. Sample containing borate and carboxylic acids

5. Selective Detectors

   (a) UV

   More selective detectors may be chosen for specific applications. Many anions (though not sulphate and phosphate) have UV spectra and such a detector has the most general application after conductivity. Choice of wavelength can make it selective for particular ions.

   (b) Amperometric

   Amperometry is very sensitive for certain species. This is especially so for sulphide and cyanide (9) neither of which can be detected by suppressed conductivity because the acids they form are not ionised.

   (c) Post Column Reaction

   Post column chemical reaction followed by colorimetric detection is the last of the more popular detectors for ion chromatography. It is used for the transition metals (10), silicate (11) and complexing anions such as EDTA (12).

6. Automation

   For even a small number of routine samples it is useful to have an automatic analyser especially when the interval between analyses is of the order of 5-10 minutes as in ion-chromatography. The automatic system used the the BP Research Centre is shown in Figure 10. The Magnus 50 position sampler controls the system and shuts it down at the end of the run. Polypropylene 2 ml sample tubes are used. These give no detectable blank for the determination of the common inorganic anions and do not require rinsing in most cases.

Ion-Chromatography in the Oil Industry

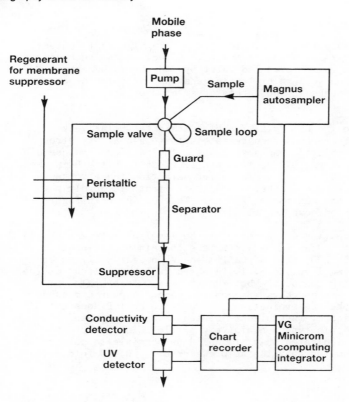

Figure 10. Automatic system based on Dionex model 10 or 2000i Shimadzu UV detector Magnus autosampler and VG Minicrom computing integrator.

Commercially, highly automated systems are available which would usually be used in specialist laboratories with a high throughput of samples such as in the water industry or the electricity supply industry.

C. APPLICATIONS

The main use of ion chromatography remains the determination of the common anions and this is true in the oil industry. Other applications such as the determination of the mono-, di- and triethanolamines from refinery gas scrubber solutions are less common. Ion chromatography can often replace a slower and less sensitive classical procedure, for example it can be used instead of gravimetry in the determination of sulphate in petroleum catalysts after extracting in the usual way with hydrochloric acid.

Two applications are discussed in more detail below. The first involves an analysis which is difficult by other means: trace sulphate in strong brines. The second is the use of ion chromatography as the detection system for elemental analysis.

1. Analysis of Produced Water

Many oil field formation waters have very high salt contents (5-20% NaCl). The determination of trace levels of anions such as nitrate and sulphate in the produced water by ion chromatography presented difficulties because of the large amount of chloride can obscure the analyte peaks in the chromatogram or overload the analytical column.

In practice the determination of nitrate was relatively simple when a UV detector was used because nitrate has a very strong UV spectrum whereas chloride has a weak one. The samples also contain bromide which has a strong UV spectrum and is close to nitrate in the chromatogram. However, by using 210 nm for detection the absorbance of bromide is greatly reduced whereas that of nitrate is hardly reduced at all from its maximum at 205 nm. Thus as shown in Figure 11 nitrate can be detected down to 0.1 µl/ml in most produced waters.

The determination of sulphate at the level of interest proved to be more difficult. Many formation waters contain significant quantities (typically 200 mg/l) of barium and when sea water, which contains about 2500 mg/l sulphate, is injected to force oil out of the well, sulphate scaling can occur when the two mix unless scale inhibitor is added. Produced waters contain other sulphur compounds so that a total sulphur determination by reduction or ICP does not give a good indication of the sulphate content. Uncontaminated formation waters of this type contain very little sulphate but published data (13) indicated that it should be possible to detect sulphate by ion chromatography in the presence of a 20,000 fold excess

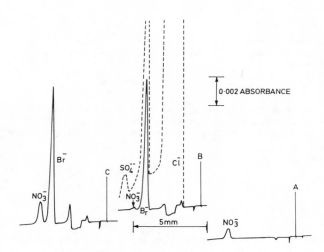

A. Injection of standard in distilled water NO_3^- 0.05 µg/ml
B. Injection of formation water diluted 1 → 20
---- Conductivity detector SO_4^{--} 0.6 µg/ml
C. Injection of formation water diluted 1 → 20
   0.125 µg/ml of NO_3^- added to the diluted solution

Figure 11. Determination of nitrate in brine.

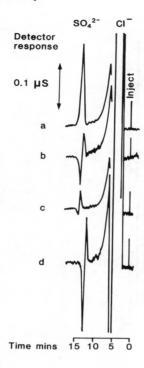

Fig 12. Chromatograms using AS3 separator column.

a) Sulphate 5 mg/l in chloride 1g/100ml
b) Chloride 1g/100ml alone.
c) Sulphate 1mg/l in chloride 1g/100ml
d) Chloride 1g/100ml, 0.1 mg/l sulphate added to

of chloride ion thus giving a detection limit of about 2 mg/l in a typical 6% sodium chloride brine. However a problem was often encountered doing this analysis. Where the sulphate peak should have been there was a small positive followed by small negative peak (Figure 12b) which prevented the detection of trace sulphate.

Investigations showed that this was due to sulphate impurity (0.01 - 0.02 mg/l) in the eluent. The effect could be increased by adding sulphate to the mobile phase (Figure 12d). By adding another ion eg oxalate to the eluent, a similar effect could be induced at the position at which that ion should elute (Figure 13). The effect is caused by overloading the separator column with chloride which displaces the sulphate on the column in equilibrium with the trace sulphate in the mobile phase. A sulphate void is left on the column which on re-equilibration in turn depletes the sulphate in the mobile phase resulting in the following negative peak.

The effect could be reduced by reducing the amount of chloride injected but this increased the effective detection limit in produced water. The level of sulphate in the mobile phase is within the limit for the analytical grade chemicals used to make it. An in-line anion exchange column was found to remove sulphate from the

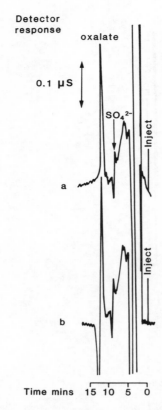

Fig 13. Eluent containing 0.1 mg/l oxalate,
AS4A separator column

a) 10 mg/l oxalate in 1g/100 ml chloride solution
b) 1g/100 ml chloride solution

eluent and eliminate the effect allowing the expected detection limit to be obtained (Figure 14).

Another means of dealing with the determination of traces in the presence of a large excess of other components is heart-cutting (14). This has been applied at the BP Research Centre to the problem of lowering still further the detection limit for sulphate in produced water. The scheme used (Figure 15) was based on a Spark Promis autosampler which has two programmable valves. After the initial separation, the peak of interest is switched to a concentrator column (switch from Figure 15a to b). The concentrator column is a short column of ion exchange material similar to the separator. It is normally used to concentrate trace ions from large sample of pure water, for example to determine the sub ng/ml levels required in the power and semiconductor industries.

For the concentrator to collect the component of interest the eluent must have been through a suppressor column. The concentrator column was then switched so that it was eluted back onto the separator column (Figure 15c). The peak of interest was separated again and

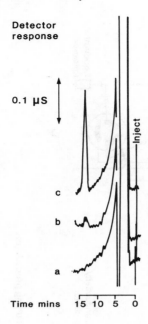

Fig 14. Determination of sulphate in 10g/100 ml chloride, AS3 separator column sulphate removed from eluent by in-line ion exchange. Sulphate: a) 0.0, b) 0.5, c) 5.0 mg/l

quantified. For the initial separation the separator column was overloaded giving a distorted sulphate peak, but collecting and reinjecting it with a much lower level of chloride sharpened it up again. A comparison is given in Figure 16 showing the effect of this process on the determination of sulphate at the 1-5 mg/l level in a 50,000 mg/l sodium chloride solution.

2. Elemental Analysis

Three standard methods of elemental analysis involve the combustion of the sample to give an aqueous solution of anions derived from the elements of interest. These are oxygen flask (IP 242 and 244), oxygen bomb (IP 61) and Wickbold oxyhydrogen burner (IP 243). Typical oil sample sizes and other characteristics of the methods are given in Table 3.

A good deal has been published about the use of ion chromatography for the oxygen flask (15) and bomb methods (16). From Table 2 it can be seen that the precision at high concentrations should be adequate. The main advantages of ion chromatography are its multi-ion capability, its sensitivity and its freedom from interference. Chromatograms obtained for the analysis of a piston deposit by oxygen flask are shown in Figure 17. The chlorine, sulphur and bromine were determined simultaneously. From the blank it can be calculated that for chlorine and sulphur, detection limits were similar to those obtained by titration ie about 0.01% whereas for bromide there was no contamination and a detection limit of 25 ppm was obtained.

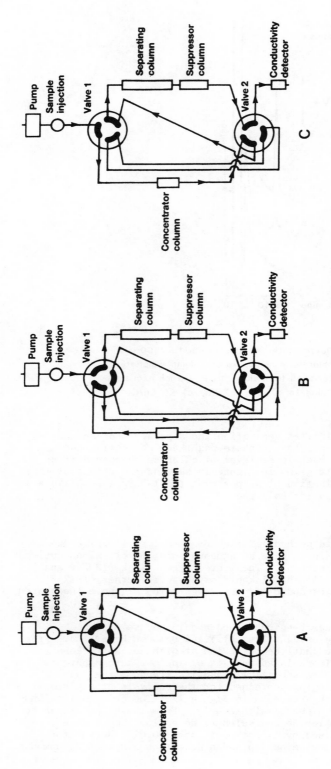

Figure 15. Valve switching sequence for the heart cutting procedure:

a) Removal of majority of chloride
b) Heart cut of sulphate peak
c) Concentrated sample reintroduced to separating columns

Ion-Chromatography in the Oil Industry

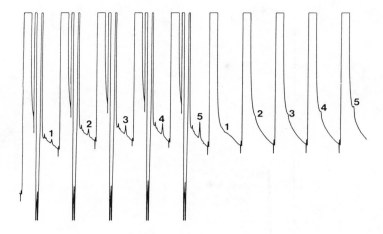

Figure 16. Sulphate in 5% NaCl solution left, with heart cutting, right, without.
(1) Na Cl alone
(2) 1 µg/ml
(3) 2 µg/ml
(4) 3 µg/ml
(5) 5 µg/ml

Detection limits using the bomb are about tenfold better than the oxygen flask for sulphur and chlorine but this is not as good as the improved sample size to final solution volume ratio would suggest because of contamination problems.

Chromatograms obtained for the combustion by Wickbold burner of a gas sample and a petroleum ether sample are shown in Figure 18. The simultaneous determination of sulphur and chlorine at single ppm concentrations is simple and detection levels well below 1 ppm are possible given reasonable care and large samples. Sub ppm levels of sulphur and chlorine are essential for the feedstocks of some of the newer catalytic refinery processes. Using the Wickbold burner and ion-chromatography these levels can be measured.

D. CONCLUSIONS AND FUTURE DEVELOPMENT

Ion chromatography is very sensitive, quite rapid, precise and is easy to automate. A number of useful applications have been given here but there is considerable scope for its increased use in refinery laboratories. It has the ability to determine sub ng/ml levels of both cations and anions. As more pure water is required for raising high pressure steam in refineries its use as a control and diagnostic tool will increase.

It has great potential in the elemental analysis of petroleum products and given sufficient interest could be added to the current

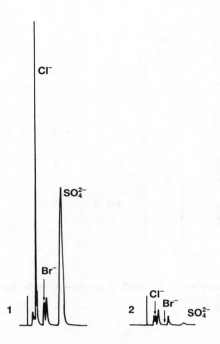

Figure 17. Oxygen flask combustions.

1. Sample containing Cl 4.3%, Br 1.8%, S 3.9%. 9.5 mg sample
2. Blank combustion

TABLE 3

COMBUSTION METHODS FOR ELEMENTAL ANALYSIS

		OXYGEN FLASK	OXYGEN BOMB	WICKBOLD
TYPICAL SAMPLE SIZE		20 mg	0.5 g	30 g
FINAL VOL ABSORPTION SOLUTION		5 ml	10 ml	100 ml
SAMPLE TYPE		solids low volatile liquids	liquids solids	liquids gases
DETECTION LIMIT WITH ION CHROMATOGRAPHY FINISH (PPM)	Cl	100	10	<0.5
	S	50	5	<0.5

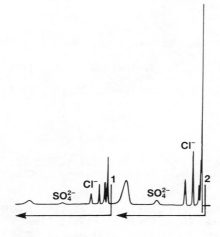

Figure 18. Wickbold combustions.

1. Petroleum gas Cl 2 ppm, S 0.5 ppm
2. Petroleum ether Cl 0.8 ppm, S 0.1 ppm

standard combustion techniques. A combination of combustion and chromatography in one commercial instrument would greatly enhance this combined technique as it could be made simple and rapid to operate.

The greatest need in ion-chromatography is for better separations especially of the anions. The clean separation given in Figure 4 does not look so good when some less common inorganic anions and numerous possible organic sulphonates are added. Improvements in column efficiency may help to improve the separation.

Another problem is that the whole range of ions which can be separated on a given analytical column can only be determined by several isocratic runs. Gradient elution would solve this problem but is only feasible using suppressed conductivity detection for UV or electrochemically inactive species. It is already beginning to be used (17) but it will require considerable development to provide the full range of separations. This is likely to consolidate the position of suppressed conductivity as the most widely used detector for ion chromatography.

E. ACKNOWLEDGEMENTS

Thanks are due to British Petroleum plc for permission to give this paper. The contributions made to the work by Dr I.M.V. Burholt, Mr T.P. Lynch and Dr J.A. Price are acknowledged.

REFERENCES

(1) H. Small, T.S. Stevens and W.C. Bauman, Anal. Chem. 47 (1975) 1801.

(2) H. Small, T.S.Stevens and C.J. Davis, Anal. Chem. 53 (1981) 1488.

(3) D.T. Gjerde, J.S. Fritz and G. Schmuckler, J. Chrom. 186 (1979) 509.
Idem ibid, 187 (1980) 35.

(4) H. Small and T.E. Miller, Anal. Chem. 54 (1982) 462.

(5) I. Molnar, H. Knauer and D. Wilk, J. Chrom. 201 (1980) 225.
M. Denkert, L. Hackzell, G. Schill and E. Sjogren, J. Chrom. 218 (1987) 31.

(6) R.M. Wheaton and W.C. Bauman, Ind. Eng. Chem. 45 (1953) 228.
G.A. Harlow and D.H. Morman, Anal. Chem. 36 (1964) 2438.
G.W. Goodman, B.C. Lewes and A.F. Taylor, Talanta 16 (1969) 807.

(7) T. Okado and T. Kuwamoto, Z Anal. Chem. 325 (1986) 683.

(8) W. Rich, F. Smith, L. McNeil and T. Sidebottom in "Ion-Chromatographic Analysis of Environmental Pollution". Eds J.P. Mulik and E. Sawicki. Ann Arbor Science.

(9) Kai Han and W.F. Kock, Anal. Chem. 59 (1987) 1016.

(10) S. Elchuk and R.M. Cassidy, Anal. Chem. 51 (1979) 1434.

(11) Dionex Application Update AU113.

(12) Dionex Application Note No 44.

(13) Dionex Appliction Note No 3.

(14) T.B. Hoover and G.D. Yager, J. Chrom. Sci. 22 (1984) 435.

(15) J.R. Krebling, F. Block, G.T. Louthan and J. DeZwaan, Microchemical J. 34 (1986) 158.

(16) R.A. Nadkarni and D.M. Pond, Anal. Chim. Acta 146 (1983) 261.

(17) Dionex I.C. Exchange 5 (1986) 1.

The Study of Lubricating Oils and Additives by Freeze-Fracture Replication Transmission Electron Microscopy—FFRTEM

K. Reading, A. Dilks and S.C. Graham
Shell Research Ltd., Thornton Research Centre, PO Box 1, Chester CH1 3SH

1. INTRODUCTION

Freeze-fracture replication transmission electron microscopy (FFRTEM) is a very powerful technique for both qualitative and quantitative characterisation of the microscopic structure of solids and liquids and as such has been used extensively in biological sciences.[1] The imaging capability of the technique gives direct visual information regarding the sizes and shapes of features and also their distribution in a solid or liquid matrix.

In this paper we present preliminary results of the application of FFRTEM to the study of laser oils, neutral and overbased detergents and dispersants which demonstrate the great potential of the FFRTEM technique in the area of lubricating oils and additives research.

This experimental programme has been pursued in collaboration with the Centre for Gell and Tissue Research, University of York.

2. EXPERIMENTAL

Since volatile specimens cannot be placed directly into the vacuum system of a transmission electron microscope (TEM), special methods of sample preparation have to be adopted. As the name implies, in the freeze-fracture replication TEM technique (FFRTEM) the volatile sample is first frozen and then fractured; the fracture surface is replicated, and it is the replica that is studied in the TEM, not the sample itself. A schematic of the specimen preparation procedure which follows is given in Figure 1. A pair of hollow, brass rivets are held together tail-to-tail in specially modified forceps, and the oil sample is introduced using a disposable hypodermic syringe with a 1 mm diameter tip (Figure 1a). The rivets and sample are then plunged rapidly into subcooled nitrogen (Figure 1b) that has previously been prepared by evacuating a vessel containing liquid nitrogen to the point of solidification. This step

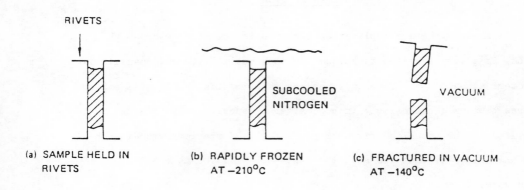

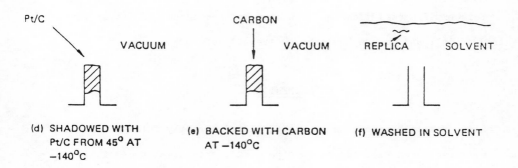

FIG. 1 — Sample preparation in the FFRTEM method

freezes the sample at a cooling rate of several thousands of degrees per second, which is usually sufficiently fast to ensure that no significant changes in the structure occur during solidification. The rivets containing the frozen sample are then transferred under liquid nitrogen to the specimen cup of a Bioetch 2005 freeze-fracture/etching/coating unit - up to three sets of rivets can be accommodated at one time in a mouse-trap-type arrangement used for subsequent fracturing. The specimen cup is then mounted, via an airlock, on the cold stage of the Bioetch unit which is held at -160°C under a vacuum of 5×10^{-4} Pa. The system's evaporators are briefly degassed prior to a short wait period, during which time the pressure recovers to a predefined operating level and the cold stage is warmed to a preset temperature, -140°C in the present work. At this point the measured pressure in the system is ca. 10^{-4} Pa, though through an intricate system of cryopumping via cold shrouds the vacuum is very much better than this close to the cold stage. Fracturing of the samples occurs when the mouse-trap-type device is triggered by an electromagnet. The jaws fly apart, separating each pair of rivets and producing a total of six fracture surfaces of the solid samples (Figure 1c). A platinum/carbon (Pt/C) evaporator is then immediately turned on, to shadow he fractured surfaces at an angle of 45° (Figure 1d). Under the conditions used, the platinum particle size in the evaporated layer is ca. 2 nm.

The surface replication is completed by the subsequent evaporation of a carbon layer in order to give the replica mechanical strength (Figure 1e).

The specimen cup is removed from the Bioetch and each rivet is unclipped and immersed in trichloroethylene in a heavy watchglass under a binocular microscope. As the sample warms and melts, the replica becomes disloged from its surface and is released into the solvent (Figure 1f). The replicas are well washed in trichloroethylene for about 2 hours with at least two changed of solvent. The clean replicas are then transferred in a

small drop of solvent into a watchglass of distilled water. The
trichloroethylene quickly evaporates and the replica spreads by surface
tension onto the surface of the water. The replicas are then transferred to
fresh water using a platinum loop and carefully picked up on a
Formvar-coated hexagonal-mesh copper TEM grid.

The procedure for sample preparation described above is generally
applicable to samples of oil-products but for samples that give highly
sculptured surfaces, the replicas can be rough on a microscopic scale and
fragment severely on washing. These small fragments of replica are useful
for qualitative examinations but are not satisfactory for deriving results
that are representative of the whole sample. The size of the replica
however, can be increased by the following additions to the experimental
procedure. After the evaporation of a carbon layer onto the frozen and
fractured surface (Figure 1e), a thick layer of silver is evaporated which
stabilizes the replica during washing . The silver is removed prior to
mounting the replica on the TEM grid by transfer to concentrated nitric acid
via a series of dilutions of nitric acid in water. Silver dissolves from
the replica after about 1 hour. The replicas are then transferred to
sulphuric acid for 1 hour and then back to distilled water via a graded
series of sulphuric acid concentrations.

The replicas are then examined in a conventional TEM with a beam
energy of 100 KeV.

During the fracturing process, the crack propagation is affected
by the hardness of the medium through which it passes. In a frozen oil
containing additive particles the direction of crack propagation may change
at each oil/particle interface where there is a change in hardness. Often,
the particle is plucked out of one fracture surface and retained in the
other. In the absence of any other changes in the fracture surface the
bumps and hollows created by such a process are visible in the
platinum-shadowed replica of the surface, studied in the TEM and have

allowed the shape and position of the particles to be imaged. However, under the conditions used here to produce the replicas, despite the fact that the samples are fractured at -140°C and then quickly shadowed with platinum and backed with carbon, there is sufficient time for a very thin layer (few nm thick) of a relatively volatile oil to sublime into the vacuum before the replication process is complete. The manifestation of this sublimation is that hollows in the fractured surface caused by plucking out very small particles are sometimes removed and the visibility of particles remaining in the surface is enhanced.

3. RESULTS AND DISCUSSION

3.1 Lubricating-oil components

3.1.1 Base oils

Since base oils themselves are complex mixtures of hydrocarbons, an understanding of their behaviour during the freeze fracture replication procedure is an important prerequisite to the study of lubricating oils and additives by the FFRTEM technique. A range of oils have therefore been studied and the TEM image of a Pt/C replicas of an oil sample is shown in Figure 2. This image is typical of all base oils we have studied and

200 nm

FIG. 2 — Base oil

exhibit a background grainy appearance due to the platinum particles in the replica and a slightly coarser texture due to the oil itself. We believe that this texture is imparted to the fracture surface by a small amount of sublimation of the lighter oil fractions into the vacuum rather than by crystallization induced by freezing. The texture is usually of low contrast (low peak-valley height) and with some samples of additive solutions and lubricating oils, is almost absent. Generally, it does not affect the observation of particles or micelles in the oil that are greater than 5 nm in size since these appear with a much higher contrast than the oil texture due to the same sublimation process (see Section 2).

When oils are cooled rapidly from room temperature with subcooled nitrogen they freeze too rapidly for the formation of wax crystals to occur. However, when oils are cooled slowly wax crystals do separate from solution. Figure 3 shows FFRTEM images of two base oils, having different pour-points, that had each been cooled relatively slowly in a deep freeze to -20°C before freezing rapidly in subcooled nitrogen. Both oils exhibit large areas in the replicas that have a similar texture to that found in replicas of the same oils frozen directly from room temperature, but in addition each show the layered structures evident in Figure 3. The layers in these structures are uniform in thickness and have extremely smooth surfaces where they have been separated by the fracture from neighbouring layers. We have assigned these structures to wax crystals in the oils and have observed them in abundance for relatively high pour-point oils frozen from -20°C. In the remainder of this present work we have eliminated the possibility of wax formation by freezing all of the additive solutions and lubricating oils from room temperature.

3.1.2 Detergents

Figures 4 and 5 show FFRTEM images of a highly overbased magnesium alkyl salicylate detergent additive in oil. The concentrate was diluted to 10%, 5% and 2.5% by blending with percolated white oil. The image in

The Study of Lubricating Oils and Additives by FFRTEM 245

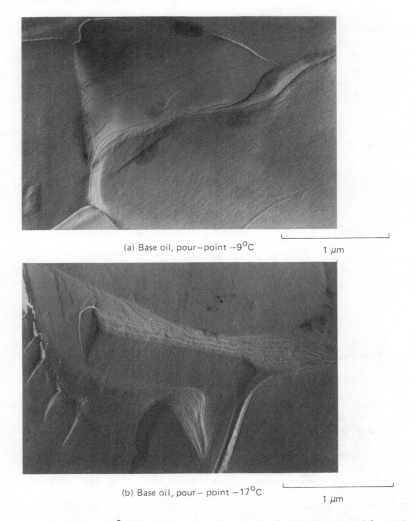

(a) Base oil, pour–point -9°C 1 μm

(b) Base oil, pour– point -17°C 1 μm

FIG. 3 — Base oil frozen from -20°C showing structures associated with wax crystal formation

Plate 4 exhibits a very rough texture when compared to base oil alone and features can be discerned that may well be additive particles. However, in such a complex image it is difficult to distinguish meaningful information from artefacts. When the concentrate is progressively diluted in going from Figures 4 to 5a and 5b, the number density of particles in the image scales closely with dilution distinguishing these particles very clearly as the additive itself.

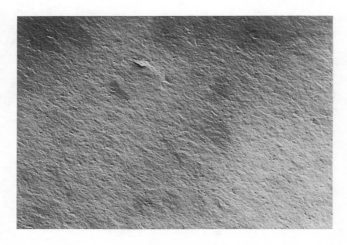

Concentrate diluted to 10% with white oil   200 μm

FIG. 4 — Highly overbased magnesium alkyl salicylate detergent

Figures 6-8 show high-magnification images of the additive concentrate diluted to 2.5% in white oil for each of three detergents; namely, a highly overbased magnesium alkyl sulphonate, a highly overbased magnesium alkyl salicylate, and a neutral magnesium alkyl salicylate. In comparing Figures 6 and 7 it can be seen that the overbased Mg alkyl salicylate is of a smaller overall particle size than the overbased Mg alkyl sulphonate even from the textures of the neat materials. The diluted samples, however, allow the particles to be measured and in principle a quantitative size distribution can be obtained. It is clear than although these overbased detergents have similar basicity levels, the alkyl sulphonate (Figure 6) consists of a low number density of relatively large particles (15-30 nm in size) and the alkyl salicylate (Figure 7) consists of a higher number density of smaller particles (5-20 nm in size).

Figure 8 shows the data for the neutral Mg alkyl salicylate. That That this detergent additive has a smaller particle size than the overbased version in Figure 7 is apparent from a comparison of the images. However, the data in Figure 8 do not show the individual particles particularly

(a) Concentrate diluted to 5% with white oil
200 nm

(b) Concentrate diluted to 2.5% with white oil
200 nm

FIG. 5 — Highly overbased magnesium alkyl salicylate detergent, showing how particle density scales with dilution

200 nm

Concentrate diluted to 2.5% with white oil

FIG. 6 — Highly overbased magnesium alkyl sulphonate detergent

200 nm

Concentrate diluted to 2.5% with white oil

FIG. 7 — Highly overbased magnesium alkyl salicylate detergent

The Study of Lubricating Oils and Additives by FFRTEM

clearly and it is therefore difficult to assess the particle size of the neutral additive. At this stage it is not known whether this is an inherent weakness in the FFRTEM technique in the study of neutral detergents or whether improved freezing and replication procedures could improve the quality of the data in Figure 8. It should be noted, however, that the dispersant imaged in Figure 9 is of a comparable particle size and is observed quite clearly.

Concentrate diluted to 2.5% with white oil   200 nm

FIG. 8 — Neutral magnesium alkyl salicylate detergent

Concentrate diluted to 15% with white oil   200 nm

FIG. 9 — Ashless dispersant 1

3.1.3 Ashless dispersants

Figure 9 shows the FFRTEM images obtained for an ashless dispersant diluted to 15% in white oil. (The magnification of the image in Figure 9 is the same as those in Figures 6-8). As was found for the detergents in the previous section; the image of the concentrate itself (not shown) shows a complex structure, but the individual particles are better seen in the diluted sample in Figure 9 where they appear as spherically shaped particles, 5-10 nm in size.

Results for another concentrate in oil are presented in Figure 10. Individual particles are difficult to discern in this image but it is clear that if micelles exist then the sizes are different. We can probably put an upper limit on their size of ca. 5 nm from the FFRTEM data.

4. CONCLUSIONS

1. The freeze-fracture replication transmission electron microscopy (FFRTEM) technique can be used to give important information relating to base oils and a variety of additives.

Neat additive concentrate

200 nm

FIG. 10 — Ashless dispersant 2

2. The sizes and shapes of detergent and dispersant particles can be determined directly from FFRTEM images for structures down to 5 nm in size. This information can be quantitative in many cases and in all cases provides a direct visual check on the validity of existing theories and measurements by non-imaging techniques.

5. ACKNOWLEDGEMENTS

Thanks are due to Dr. A.J. Wilson at the Centre for Gull & Tissue Research at the University of York without whom this work would not have been possible.

6. REFERENCE

1. Replica, shadowing and freeze-fracture techniques, J.H. Wilson and A.E. Lowe. R.M.S. 'Practical Methods in Electron Microscopy' Series North Holland, 1980.

Petroanalysis '87
Edited by G. B. Crump
© 1988 John Wiley & Sons Ltd

The Analysis of Polychlorinated Biphenyls (PCBs) in Mineral Based Insulating Oils by Capillary GLC

K.J. Douglas and R.A. Pizzey
Castrol Research Laboratories, Pangbourne, Reading RG8 7QR

Summary

This paper describes the development of an analytical method for the analysis of polychlorinated biphenyls (PCBs) in mineral based insulating oils using high-resolution capillary gas-liquid chromatography (GLC) and detection by electron-capture detector (ECD), with the use of a peak identification computer program for data processing.

It will also deal briefly with some background on PCBs.

What are PCBs?

A PCB is formed by the addition of one to ten chlorine atoms onto a biphenyl ring system resulting in a possible 209 individual PCBs, more correctly known as congeners, (including monochlorobiphenyls).

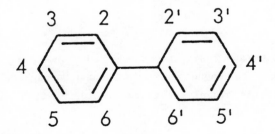

Figure 1

They have been produced as commercial mixtures by a variety of manufacturers from around the world via the direct chlorination of biphenyl with anhydrous chlorine. PCBs are not known to occur naturally, and although 209 congeners are theoretically possible, some are statistically unlikely and have not been observed in commercial mixtures.

PCBs can vary from one manufacturer to another and from batch to batch in terms of the individual congener concentrations within a particular grade of a commercial mixture. However, some commercial products, notably those produced by Monsanto and known under the trade name Aroclor, are frequently used as reference standards, e.g. ASTM D4059/86 - a packed column GLC method. Most information concerning PCBs is related to the Aroclors which are characterised by a four digit number. The first two digits represent the type of molecule with 12-- denoting a chlorinated biphenyl mixture and the second pair of digits denoting the percentage (w/w%) of chlorine present in the product (e.g Aroclor 1260 is a mixture of PCB congeners with 60% Cl).

Properties and Uses

PCBs are thermally stable and inert, resisting oxidation, acids, bases and other chemical agents. They have excellent dielectric properties and are compatible with organic materials as well as being nonflammable. This led them to have many industrial applications. Their more recent uses include electrical capacitors, electrical transformers, vacuum pumps and gas-transmission turbines. Currently the use of PCBs in many applications is banned and is strictly controlled in others, depending upon the country concerned. Large scale industrial production has ceased.

Why do PCBs cause Concern?

Due to their lipid solubility and resistance to degradation, along with uncontrolled disposal, PCBs are found as contaminants throughout the world in many components of the global ecosystem, including fish, birds and man.

In the 1960s, PCBs were seen as unknown peaks during the analysis of pesticides and in 1966 their widespread occurrence in the Swedish environment was first noted.[1] Since this time, numerous studies relating to PCBs have been made, many relating to their analysis (which is complicated by the large number of matrices in which they are found).

Analysis of PCBs in Mineral Based Insulating Oils

The toxicity of PCBs is a matter of continuing debate. Early studies in the United States showed PCBs to cause cancer in animals while a report in 1981 by the National Institute for Occupational Safety and Health[2] produced a mortality study of 2,500 PCB-exposed electrical workers, over 50% of whom were exposed to PCBs on the job for over 20 years, with some up to 40 years. The results showed no significant excess in total mortality from cancer, cardiovascular disease or any other cause.

Despite this lack of evidence from about 50 years of commercial use, human health risks continue to be studied due to the widespread contamination by PCBs and their continued input into the environment.

Requirements for Analysis and Choice of Instrumentation

A commercial requirement arose to have an analytical system that could initially analyse PCBs according to ASTM D4059 using packed column GLC and ultimately to analyse the individual congeners using state-of-the-art GLC technology. Automation was desired to reduce operator time and facilitate overnight analysis. Automated, cold on-column injection to provide the most accurate quantitative analysis would have been ideal but was not available at the time. Instead, a split-splitless injection system was chosen with an autosampler, the GC being compatible with existing instruments in terms of servicing and spares.

A detector specific for PCBs was required to eliminate response from the mineral oil. An electron-capture detector (ECD) and a Hall Electrolytic Conductivity Detector (HECD) were considered. The HECD is halogen specific, and responds only to the amount of chlorine present in a molecule so sample clean-up is often not needed. The ECD is less specific thus sample clean-up is frequently required. It also responds differently to the various sample components such that for PCBs the response factors (RFs) vary by a factor of over 20. However, the ECD has been reported to be easier to use, maintain and troubleshoot with better precision and greater sensitivity, whereas the HECD has given problems with coking and column overload. The ECD was considered to be most suited for the application.

The autosampler chosen had advantages of being compact, easy to use, and having an efficient flushing system whereby the sample is pushed rather than sucked through the syringe assembly.

The best possible separation of the individual congeners within a reasonable time led to the selection of a high efficiency, narrow-bore capillary column 50 metres in length.

The ECD is operated at 350°C to stop condensation of the eluting components. Any air or water present degrades the ECD performance so all gases passing through the detector have to be filtered through both oxygen and moisture traps.

Methodology and Early Development

When development work commenced there was only incomplete data available on the response factors of different congeners in the ECD. It was decided to derive these factors for those peaks which could be obtained from commercially available PCB mixtures. The principle was to split the column effluent to both the ECD and FID, quantify the PCB mixtures using their equi-molar responses from the FID and so deduce response factors for the individual congeners for the more sensitive ECD. This approach had been reported in 1981.[3] Considerable problems were encountered in practice due to the large difference in response between ECD and FID to the PCB congeners. This approach was abandoned when full response factor data was published, as described below.

Splitless injection was tried with the 'recondensation effect', where the sample (+ solvent) vapour is condensed at the front of the column by the use of an oven temperature below that of the solvent boiling point. This causes a narrowing of the initial bands and results in sharper eluting peaks and thus improved chromatography. However, condensation of septum bleed and its subsequent elution as spurious peaks proved a great problem at high ECD sensitivities, (even with septum purge in use). To overcome this problem a higher initial oven temperature was used. A small split was needed to improve sample transfer as the recondensation effect could no longer be utilised. This solution also provided a faster cool-down time from the final to initial oven temperature.

Change of Plan and Final Method Development

An excellent paper published in 1984[4] caused a change of plan with the saving of many months of work. In this work all 209 PCB congeners had been synthesised and their individual response factors and retention times determined relative to an internal standard.

Use of a similar type of column enabled these data to be adapted to determine the elution order of the individual congeners and thus the relative retention times (RRTs) to the internal standard (decachlorobiphenyl) on this GC system. This was done after first optimising the chromatography in terms of carrier gas flow-rate, oven temperature-program, auto-sampler parameters, etc. using a carefully chosen mixture of commercial products that included all the PCB congeners likely to be observed in the samples.

A slow ramp-rate of 1°C/minute from 100°C to 290°C produced a well resolved chromatogram that was used for peak assignment but too long (3 hours) for routine use. Other temperature programs were tried in order to obtain a shorter analysis time with no significant loss in resolution compared to the original chromatogram. This was achieved with a ramp-rate of 2.5°C/min from 130°C to 290°C, producing 98 individual peaks. (See Figure 2.)

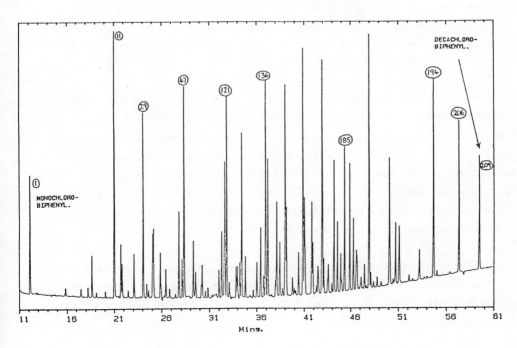

Figure 2

Sample clean-up

Mineral oil causes a depression of the ECD signal; however, this is largely overcome by dilution of the sample to 1%(w/w) in solvent, and all standards are similarly prepared. Interferences, such as oxidation products must be removed as much as possible. There are many methods for this in the literature; some are quite complex and others require further expense on automation. It was found that a simple, manual clean-up procedure was adequate for the needs of this method. It involves shaking the diluted sample with half its volume of concentrated H_2SO_4, leaving to stand, then shaking the solvent decanted layer with 'Florisil' adsorbent previously activated at 130°C.

Data Processing

The data processing is important to the success of this method. It involves the use of a DEC PDP11 mini computer with 'in-house' software. The raw data is captured and converted to peak data. The internal standard (decachlorobiphenyl, DCB) is given a RRT=1 and then all other peaks have RRTs compared to that. These RRTs are compared to those on file of all the possible congeners. Any peaks outside the RRT by ±0.0015 are disregarded such that any remaining non-PCB peaks are eliminated. The appropriate RFs are used on each identified peak to calculate the concentrations of the individual PCBs (or groups thereof, if coeluting) and then summed to produce the total PCB content.

The RFs have been weighted for known mixtures of PCBs (eg. Aroclor 1260) from published data,[3,5] depending on the relative proportions of the individual congeners that may coelute under one peak. The most appropriate file of RFs may be chosen depending upon the knowledge of the sample. Files of RFs have been produced for 'all possible congeners', 'all probable congeners', and Aroclors 1248, 1254, and 1260 (See Figure 3). In practice the 'all probable' file is the most useful especially with unknowns and mixtures of commercial products.

A printout is obtainable showing the individual congener(s) concentration against the peak number from the run along with the designated congener numbers (Ballschmizer and Zell)[5] and full names. Simplified reports with less information are also available.

Peak Number	RRT	Congener Number	All Possible	All Probable	RESPONSE FACTORS (RFs)		
					1248	1254	1260
34	.51264	70, 76	.6207	.6188	.658	.650	.650
35	.51862	66, 80 93, 95, 102	.6050	.5881	.646	.646	.646
36	.52225	121	.766	.766	.766	.766	.766
37	.52705	55, 91	.7000	.7000	.586	.591	.580

Figure 3 - Extract from Data Files

Conclusions

The high resolution obtained by capillary GLC has enabled the development of a method that can deal with mineral-based insulating oil containing PCBs from any source and produce a reasonable estimate of the total PCB content or the concentration of an individual congener if needed. This would not be possible with low resolution packed column GLC where interferences can easily be hidden and mixtures of commercial products are difficult to quantify.

Acknowledgement

The authors thank Castrol Ltd. for the use of Company information and permission to publish.

References

[1] = M. D. Erickson, "Analytical Chemistry of PCBs", Ann Arbor Science (Butterworth Publishers, 1986).

[2] = D. P. Brown and M. Jones, Arch. Environ. Health, 1981, 36, 120-9.

[3] = P. W. Albro et al, J. Chromatog. 205 (1981) 103-111.
[4] = S. H. Safe et al, Environ. Sci. Technol. 18 (1984) 468-476.
[5] = K. Ballschmizer and M. Zell, Fresenius Z. Anal. Chem. 302 (1980) 20-31.

APPENDIX

Instrument Details

Sigma 300 Gas Chromatograph with ECD, FID, and AS 300 autosampler (Perkin-Elmer).

DEC PDP11 minicomputer

Column; 50m x 0.2mm id x 0.11um   5% phenyl methyl silicone an "Ultra Performance" capillary column (19091B OPT.005, Hewlett-Packard).

GLC Conditions

Helium carrier @ 41psi   (1ml/min @ 130°C)
Injector temperature; 270°C
Detector temperature; 350°C
ECD; Range 1   95% argon/5% methane @ 60ml/min
Oven temperature program; 130°C —————— 290°C
                         0 mins   2.5°C/min   0 mins
Split 4ml/min
Autosampler; Injection volume 3ul
40 second flush with He @ 7psi
2 second injection delay

Sample Preparation

Iso-octane containing 0.02ppm DCB used as diluent.

1% sample in diluent (~2-3ml total), shaken with half its volume of concentrated H_2SO_4 and left to stand. Iso-octane layer removed and shaken with 'Florisil' previously activated at 130°C. Left to settle, then analysed.

The Application of Chemometrics to ^{13}C NMR Spectra of Hydrocarbon Mixtures

A.G. King* and J.M. Deane
* Esso Research Centre, Milton Hill, Abingdon, Oxon OX13 6AE
AFRC Institute of Food Research, Bristol,

Introduction

An example of the application of Chemometric procedures to the analysis of a complex hydrocarbon mixture of lubricating oil basestock components is described below.

The development of more advanced passenger car engines has resulted in the need for increased performance from the engine lubricating oils. This can be achieved by using blends of several hydrocarbon components rather than just a single mineral oil basestock. Thus, the oil portion of a modern engine oil can comprise mixtures of mineral oil, (MO), hydrotreated petroleum stock, (HC) and polyalphaolefin materials (PAO).

Due to their different properties and large cost differentials (Table I) it is important that they can be estimated fairly accurately in both test blends and finished products.

Table I

Some Properties of Lubricant Components

	PAO	HC	MO
MW Range	300-500	200-500	200-700
Pour Point	-70C	-15C	-20C
Viscosity	6 cSt	6 cSt	6 cSt
Volatility	2%	10%	7%
VI	140	150	100
Cost	8X	4X	X

Several papers,[1,2,3,4] have dealt with the properties of these component types.

Idealised structures for these components are shown in Figure 1. Commonly, the R group is C_7 and therefore the starting alphaolefin is decene. It should be noted that only the trimer is shown whereas the material will comprise dimer, trimer, tetramer, pentamer in various proportions depending upon the desired viscosity.

In the case of the petroleum it should be remembered that there are numerous carbon isomers possible. Also, not shown, is the aromatic portion of the structure which may account for 1-8% of the total. The structure of the hydroprocessed component will be similar to that of the petroleum except that the aromatics will be considerably reduced and the average molecular size will be reduced with consequent increase in branchiness.

POLYALPHAOLEFIN

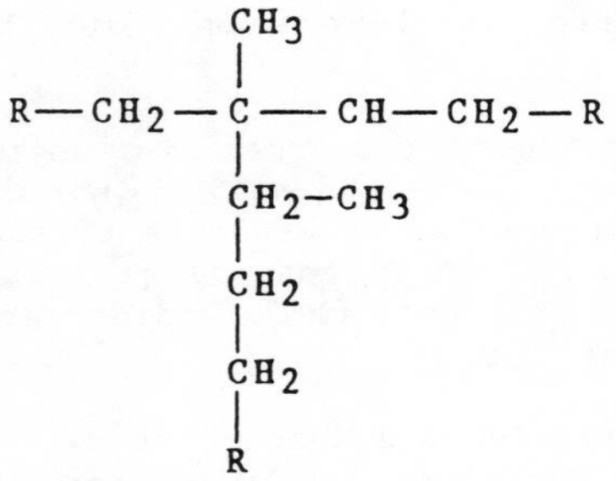

$$R-CH_2-\underset{\underset{\underset{\underset{R}{|}}{CH_2}}{\underset{|}{CH_2}}}{\overset{\overset{\overset{CH_3}{|}}{|}}{\underset{|}{C}}}-CH-CH_2-R$$

with side chain CH_2-CH_3 on the central carbon.

Trimer

PARAFFINIC HYDROCARBON

(PETROLEUM)

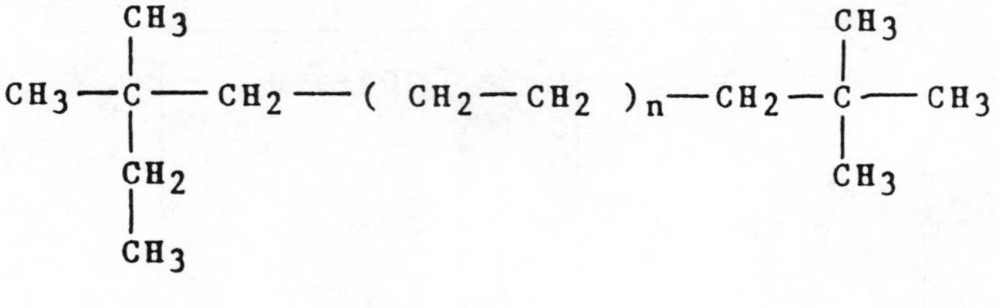

$$CH_3-\underset{\underset{CH_3}{|}}{\overset{\overset{CH_3}{|}}{\underset{|}{C}}}-CH_2-(CH_2-CH_2)_n-CH_2-\underset{\underset{CH_3}{|}}{\overset{\overset{CH_3}{|}}{\underset{|}{C}}}-CH_3$$

with CH_2 branch on the left quaternary carbon.

Where n = 3 to 10 units

Figure 1

From even these brief descriptions it will be readily appreciated that these three components are all very similar in chemical composition and this plus the fact that they are each mixtures in their own right makes their separation extremely difficult.

Conventional analytical separation methods such as chromatography do not enable a separation of these materials to be made. The current Esso Research Centre, Abingdon scheme for the separation of a fully formulated lubricating oils is shown in Figure 2 and serves to illustrate this point. You will notice that the basestock components are all found in the same portion of the diagram and the subsequent identification and quantification (but not separation) is acheived by MS and CNMR.

The MS analysis is necessary since, although we are not yet able to quantify the components by MS, they can be identified readily by this technique whereas the CNMR analysis assumes the presence of the components.

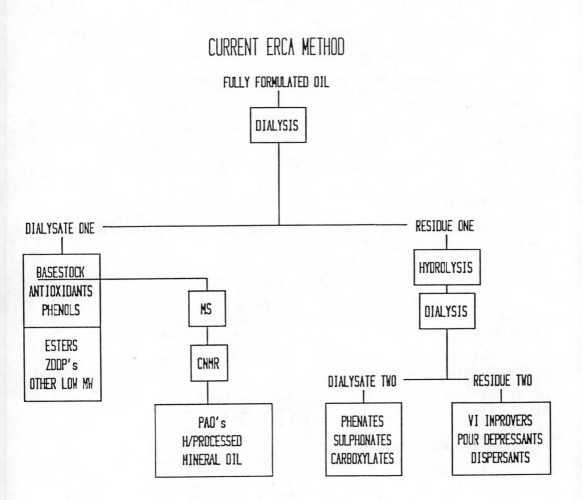

Figure 2

Figures 3 through to 6 show the ^{13}C NMR spectra of the aliphatic regions of each of the three types of component plus the spectrum of a 1.5:1:1.5 mixture of the three. It will be appreciated from these that the analysis of an unknown mixture from its ^{13}C spectrum is a far from trivial process.

Application of Chemometric Procedures

Consideration of the spectra shows that there is a vast amount of data available in each one. With the conventional method of making up known linear blends and then analysing all the results the final data analysis is very difficult and requires a high level of NMR interpretative skill.

On the other hand, the high data content would seem to be ideal for the chemometric approach to the problem. Thus it was decided to attempt this analysis using chemometic techniques.

From consideration of other analytical [5,6,7,8] studies it was decided to apply the techniques of Principal Component Analysis/Canonical Variate Analysis (PCA/CVA) with Target Rotation and Least Squares Regression. Detailed discussions of these techniques and examples of their application to the pyrolysis-mass spectra are to be found in the references.

For the three component mixture in this work the experimental construction was based on a {3,4} Simplex Lattice Design, giving an 18 point mixture design as shown in Figure 7.

Twelve samples were made up in duplicate and run as replicates, namely numbers 1 through to 8 and 10,11,12 and 17.

The composition of the samples can be found in Table IV.

Experimental

The ^{13}C NMR spectra were obtained on a JEOL FX90Q FT spectrometer operating under the conditions detailed in Table II.

Table II

Instrument Conditions

Solvent	Deuterochloroform
Lock	Deuterium
Relaxation Agent	Approximately 20mg/ml Cr(AcAc)$_3$
Pulse Angle	30°
Pulse Delay	2s
Memory Size	For digitisation rate >1.2 Hz
Spectral Width	250 ppm
Filter Width	250 ppm
Decoupling	Broad Band Gated
Decoupler Rise Time	<2ms
S/N	Typically 60:1 for CDCl$_3$
Line Broadening	1* Digital Resolution
Spin Rate	Approximately 25Hz
Temperature	50° C
Plot	Aliphatic Region only, 0-60 ppm
Reference Peak Position	14.2 ppm

POLYALPHAOLEFIN

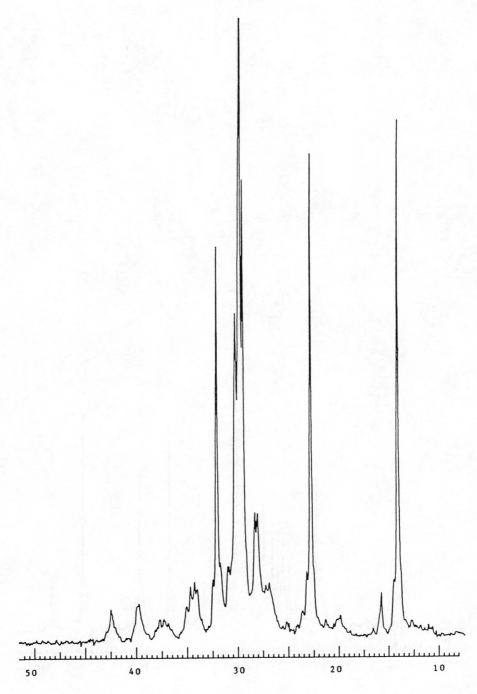

Figure 3

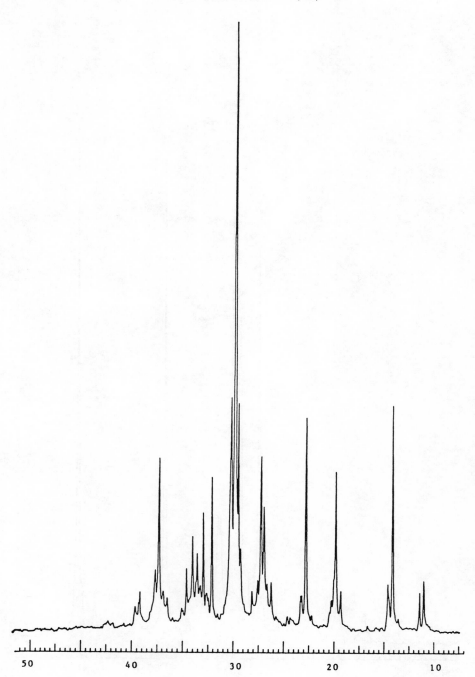

Figure 4

Application of Chemometrics to ^{13}C NMR Spectra of Hydrocarbon

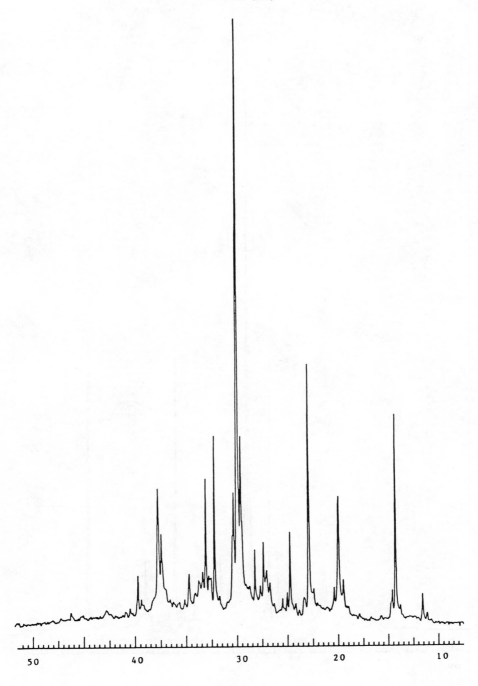

Figure 5

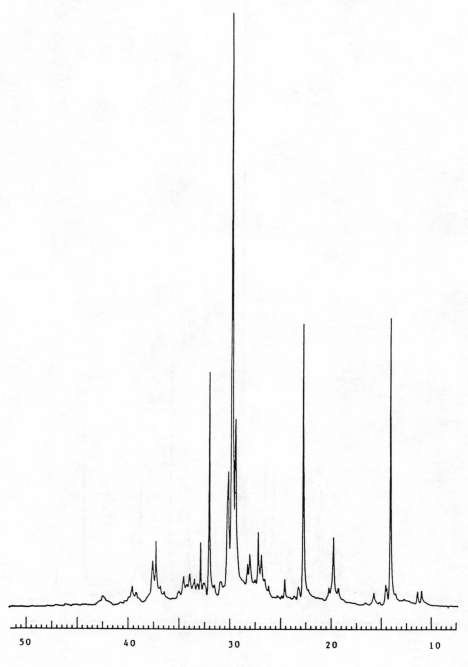

Figure 6

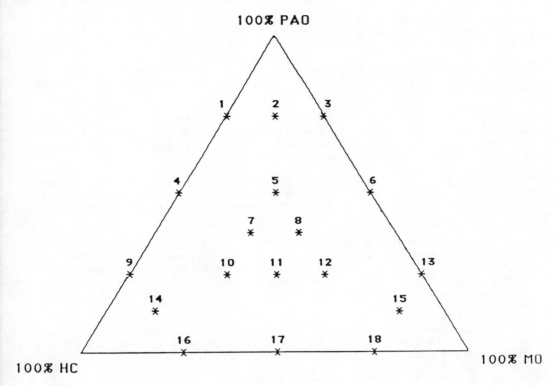

MIXTURE TRIANGLE SHOWING THE PROPORTION OF CONSTITUENTS FOR 18 EXPERIMENTAL MIXTURES

Figure 7

Results and Discussion

A number of different statistical techniques were applied to the data set following the lines of Vallis et al [5,6].

Initially PCA and CVA were applied to the data set to acquire a general feel for the data.

PCA on the original data indicated a good distinction of PAO proportions. Purer HC and MO were distinct, but blends of the two were skewed to HC except for point 15 which appeared to be an outlier and was also an outlier in the third dimension indicating that this sample had some peaks distinct from all other samples.

Data analysis was repeated with a replicate for position 15 and again PAO proportion was distinct with HC and MO blends skewed as before.

The replicate of point 15 followed the pattern for all the other blends but it was also an outlier in the third dimension. This indicated that the replicate samples of point 15 have peaks distinct from each other and all other samples.

CVBA was then applied to the data and again a plot of the first two CV's recreated the mixture triangle. PAO distinction was good but MO/HC blends were still skewed.

When either of the samples for position 15 were included in the analysis this point was an outlier. However, when both were included it behaved like all other blends, indicating that the within sample variation of replicates is more dominant than individual peak height differences.

PCA and CVA analysis show good evidence of reproduction of the mixture triangle from the peak heights of the blends however, the results indicate that a correction factor may be required to exactly predict the proportions of HC and MO.

The data was then analysed by PCA and CVA followed by Target Transformation and component proportion estimation. In this analysis the pure components are not included in the analysis but are projected onto the mixture space after analysis. This information is then used to predict the component proportion under the assumption that all blends are a linear combination of the pure components.

The PCA analysis showed position 15 to be an outlier when either or both replicates were included in the data, resulting in a poor prediction of the blend composition.

When the samples for position 15 were removed from the analysis the results obtained were greatly improved PAO discrimination was still evident but now HC/MO blends were skewed towards MO.

This change in skewness and greater outlying position for position 15 can only be caused by the removal of the pure constituents. This indicates that blends primarily made up from MO with little HC content are distinct from other blends.

The CVA analysis was again dominated by position 15 when either of its replicates was included in the analysis. When both samples for position 15 was included the results obtained gave a very poor reproduction of the mixture triangle. When position 15 was removed the mixture triangle and the component estimates obtained were the best results obtained thus far. The triangle showed no particular bias in any direction for HC/MO blends but the component estimation was still not accurate enough to use as a predictive tool.

The data was reanalysed by CVA of the dominant PC's, and by doing this the PCA removes the random variation in the data prior to CVA.

When any of the replicates for position 15 were included in the analysis the results were dominated by this position, since this position was an outlier in the second and third dimensions in PCA. When position 15 was removed the results obtained were even better than CVA followed by Target Transformation. The worst estimate is for position 6, which again is a blend with a high MO content and no HC.

In summary the best method to recreate the mixture triangle and estimate the component proportions is by carrying out a CVA followed by Target Transformation on the dominant PC's of the data. The different analyses have shown problems in the area of high MO content blends.

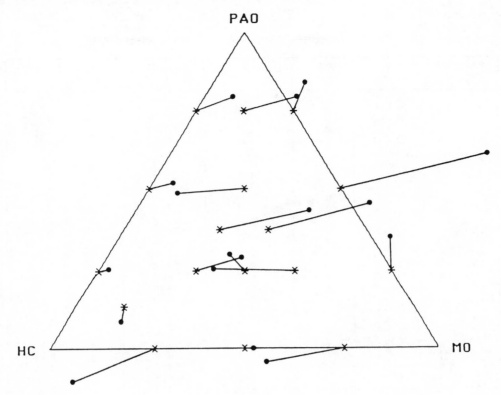

KNOWN AND ESTIMATED COMPOSITION OF 17 MIXTURES OF PAO, HC AND MO. THE ASTERISKS DENOTE THE KNOWN COMPOSITION AND THE DOTS THE ESTIMATED COMPOSITION.

Figure 8

Figure 8 shows the mixture triangle with the actual and estimated composition of 17 blends, obtained by applying PCA/CVA with Target Rotation and Least Squares Regression to a date set of 29 samples analysed by ^{13}C NMR.

One blend, Number 15, was removed from the analysis due to outlier peaks in its spectrum which distorted the results.

A total of 58 peaks, which contained about 95% of the spectral information was obtained from each sample.

Example data for the 100% pure components and the 1.5:1:1.5 mixture are shown in Table III.

Table III

Distribution and Composition of Selected Blends

Peak No.	Chemical Shift	100% PAO	100% HC	100% MO	37.5:25:37.5 Mixture
1	11.1	.3	1.3	.4	.6
2	11.5	0	.9	.9	.6
3	13.7	0	1.1	.7	.5
4	14.0	1.4	.7	0	2.1
5	14.2	10.8	5.3	5.4	8.5
6	14.5	1.7	0	1.9	1.5
7	14.7	0	1.2	0	.8
8	15.8	1.1	0	0	.5
9	19.4	0	.9	1.3	.7
10	19.6	0	.5	0	0
11	19.9	.5	3.6	3.6	2.4
12	20.3	0	.8	1.1	.7
13	20.5	0	.6	0	0
14	22.4	.4	0	1.1	0
15	22.6	1.3	0	1.0	0
16	22.9	10.2	6.0	6.7	7.6
17	23.3	1.6	.9	1.9	1.6
18	23.7	.7	0	0	.4
19	24.1	.4	0	.7	0
20	24.7	0	.4	2.6	1.0
21	25.0	0	0	1.0	.5
22	25.4	.4	0	.8	.5
23	26.3	0	1.2	0	.8
24	26.7	0	1.2	1.2	1.0
25	27.0	1.2	3.0	1.6	1.9
26	27.3	1.1	4.3	2.3	2.2
27	27.6	0	1.3	1.2	1.0
28	28.1	2.6	1.0	2.0	1.7
29	28.4	2.7	0	0.9	1.6
30	28.8	0	.5	2.3	1.0
31	29.3	0	2.0	0	0
32	29.6	9.9	4.9	4.8	7.5
33	29.9	17.7	21.8	19.0	18.7
34	30.3	7.7	7.9	3.9	4.5
35	31.0	3.8	.7	.9	1.0
36	31.8	1.7	.5	1.0	1.7
37	32.1	9.0	4.3	4.8	8.1
38	32.6	1.9	1.0	1.5	1.8
39	33.0	0	2.7	4.0	2.1
40	33.2	1.0	1.1	1.7	.9
41	33.7	0	3.1	1.4	1.0
42	34.0	1.3	2.2	1.1	1.3
43	34.4	1.5	0	0	.9
44	34.6	0	1.5	1.6	1.2
45	34.8	1.4	0	0	0
46	35.1	1.0	.6	0	.6
47	35.6	0	0	.8	0
48	36.2	0	0	.8	0
49	36.5	0	.8	0	.6
50	36.9	.6	1.0	0	.8
51	37.4	.6	4.2	2.5	1.9
52	37.7	.6	1.4	4.1	1.8
53	39.3	0	.9	.9	.6
54	39.7	0	.7	1.5	.8
55	39.9	0	0	0	0
56	40.4	1.0	0	.6	0
57	42.6	.9	0	0	.5
58	46.2	0	0	.5	0

It should be noted that the peak concentrations are reported on the basis of normalised peak heights rather than the usual peak areas. This approach is entirely based on the ease with which such data could be obtained from the spectrometer as opposed to the difficulty in obtaining integrated areas for all 58 peaks.

Figures 9 through 11 show the prediction errors of all 17 blends for each component while Table IV shows the actual blends, estimated blends and prediction errors.

Table IV

Actual and Estimated Composition of 17 Mixtures of PAO, HC, and MO Component Concentrations (%)

Blend	Actual			Estimated			Difference		
	PAO	HC	MO	PAO	HC	MO	PAO	HC	MO
1	75	25	0	79.6	12.7	7.7	4.6	-12.3	7.7
2	75	12.5	12.5	79.6	-3.3	23.7	4.6	-15.8	11.2
3	75	0	25	83.9	-7.9	24.0	8.9	-7.9	-1.0
4	50	50	0	52.3	41.5	6.2	2.3	-8.5	6.2
5	50	25	25	48.7	42.8	8.4	-1.3	17.8	-16.6
6	50	0	50	60.9	-43.7	82.8	10.9	-43.7	32.8
7	37.5	37.5	25	42.7	11.7	45.6	5.2	-25.8	20.6
8	37.5	25	37.5	45.4	-5.5	60.1	7.9	-30.5	22.6
9	25	75	0	25.8	73.2	1.0	0.8	-1.8	1.0
10	25	50	25	28.8	37.5	33.6	3.8	-12.5	8.6
11	25	37.5	37.5	29.8	39.5	30.7	4.8	2.0	-6.8
12	25	25	50	25.1	45.6	29.3	0.1	20.6	-20.7
13	25	0	75	34.8	-5.0	70.2	9.8	-5.0	-4.8
14	12.5	75	12.5	8.4	77.6	14.0	-4.1	2.6	1.5
16	0	75	25	-9.7	99.5	10.2	-9.7	24.5	-14.8
17	0	50	50	0	48.5	51.5	0	-1.5	1.5
18	0	25	75	-3.2	46.7	56.5	-3.2	21.7	-18.5

ESTIMATES OF MO PROPORTIONS AGAINST TRUE PROPORTIONS

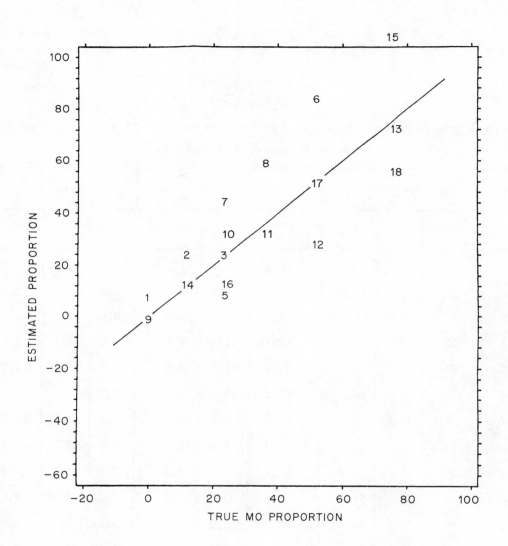

POINTS COINCIDING WITH POINT
4

Figure 9

ESTIMATES OF HC PROPORTIONS AGAINST TRUE PROPORTIONS

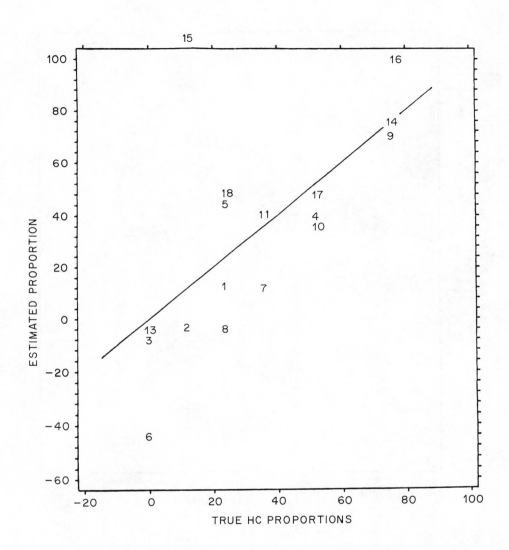

POINTS COINCIDING WITH POINT 5
12

Figure 10

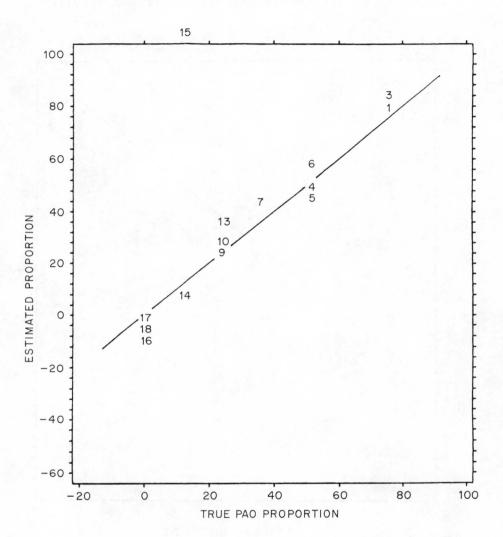

Figure 11

Overall, the estimate of PAO content of the blends is of the order of 10% absolute and is therefore of acceptable accuracy for analytical purposes. However, HC and MO estimates are much less satisfactory with the HC component up to 44% misestimated in one case and the MO component 33% misestimated on the same sample.

From a knowledge of the NMR procedures it is thought that the problem may lie in the data presentation rather than any inherent defect in the statistical techniques applied to the problem.

Future work will address this aspect of the analysis.

Conclusions

The feasibility of using statistical techniques to estimate the properties of components in a complex three component mixture has been demonstrated.

In doing so it has been shown that the total data from a ^{13}C NMR spectrum can be used without expert interpretative skills to produce an analytical result.

Since no interpretation is involved in the analysis the time taken to obtain an analytical result is shorter than by conventional calculation.

Finally, it can be seen that there is much information to be gained about the interactions of the components through analysis of the information provided by chemometric techniques.

References

1. T. Suzuki et al., J. Japan Petrol. Inst. 24, 3, 151 (1981)
2. J.B.Boylan & J.E.Davis, J. Am. Soc. Lub. Eng. 40, 7, 427 (1984)
3. J.A.Brennen, ACS Washington Meeting, September 14th 1979
4. R.L.Shubkin et al., ibid.
5. L.V.Vallis et al., Anal. Chem. 57, 3, 704 (1985)
6. L.V.Vallis et al., J. Anal. & App. Pyrol. 5 333 (1983)
7. R.A.Hearmon et al., J. Chemo. & Intell. Lab. Systems, 1, 2, (1986)
8. O.M.Kvalheim et al.,Anal. Chem. 57, 14, 2858 (1985)

Tough Problems—Novel Solutions

D. Betteridge
BP Research Centre, Sunbury-on-Thames, Middlesex, TW14 7LN

The analytical laboratory at the Research Centre, Sunbury addresses significant analytical problems by a variety of methods. In this lecture three case histories will be discussed to demonstrate how novel methods and approaches can be used to solve some difficult problems. In particular, emphasis will be given to the impact of computers and artificial intelligence.

Catalyst Theta 1

Zeolite catalysts are of great importance to the oil industry. The problem put to the Analytical Division via the development of the Zeolite Theta 1 was how, if one is limited to powder samples, can one make a structural determination which will stand up in a patent case? This was answered by a combination of techniques. First some idea of the possible structure was obtained from powder X-ray diffraction and NMR and this was supported by solid state NMR. Both physical and computer graphic models were made. The computer modelling and molecular graphics permitted variations in structure to be simulated and were used to enhance the interpretation of the X-ray diffraction and NMR and to predict the electron microscopic image. This last was done by compounding the model with the known electron optical aberration to give a computer graphic projection of what would be observed under the electron microscope. The projected computer image and the microscopically matched thus confirming the structure.

HPLC Scanning of Plates

There is considerable interest in the possibility of using HPLC as a quantitative method, but this is limited by the practical difficulty and time of measurement of spots with a standard optical mechanical device. A simple solution is to use a video camera to observe the whole plate and standard image processing software to detect and measure the spots. This can be completed within a few minutes and reduces the analysis time of a plate by several hours. The advantage of the speed is illustrated with spots whose intensity is decaying and which cannot reliably be measured by conventional means. These are easily catered for by the video approach where the time of measurement is small and the whole plate is virtually measured simultaneously so that relevant intensities are readily determined. Reasonable correlations have been found between concentration and peak areas as determined by this method.

Analytical Trouble-Shooting and Expert Systems

The preservation of expertise and enhancement of an expert's performance are long standing problems which today are, in many organisations, exacerbated by the age of existing experts and the reduction of manpower. Expert systems in which the espertise can be retained and accessed offer an interesting and potentially valuable solution to this problem.

We explored the potential with a test case which was to aid an analytical expert trouble-shoot on the performance of a gas dehydration plant. This is a problem which is faced about twice a year and involves the expert going to the plant, taking a true sample of the gas, carrying out an independent analysis, performing a calculation of plant performance from chemical engineering principles, comparing the analytical results with those projected from the calculations and making recommendations for remedial action. The expert system device enables the expert to do preliminary work on P&I diagrams to identify suitable points at which reliable samples may be taken. It then checks the analysis, performs the chemical engineering calculations and provides the expert on site with advice on the interpretation of the results.

A lot of hype has been generated over expert systems and it is important to bear in mind that they are only computer programs. The results obtained with our system have encouraged us to continue to develop other systems. Some of the practical difficulties such as the acquisition and representation of knowledge and the validation of the system will be described.

Developments in the Analysis of Petroleum Products by Capillary Chromatography

C.A. Cramers
Laboratory of Instrumental Analysis, Eindhoven University of Technology, POB 513, 5600 MB Eindhoven, The Netherlands

Abstract

The presentation deals with recent trends in mainly open tubular chromatography of importance to petroanalysis.

Part 1

Based on dimensionless parameters equations are given which compare speed of analysis, pressure drop and plates per bar pressure drop for the three forms of fluid chromatography: GC, SFC and LC. The treatment includes the specific advantages of capillary and packed columns in special applications. If extreme speed of analysis is strived for, contemporary packed columns are superior to open tubulars, in SFC and LC. Capillary columns are the columns of choice, if high plate numbers are needed and pressure drops is limiting factor (like in SFC and GC). Some state of the art examples of separations in the three forms of capillary fluid chromatography will be presented.

Part 2

Part 2 concentrates on recent developments in GC: narrow bore thin film columns (typically $50 \mu m$ inside diameter; $0.05 \mu m$ film thickness; wide bore thick film columns ($500 \mu m$; $5-10 \mu m$ bonded phase films); PLOT columns using adsorbents as the stationary phase.

Variation of the column inner diameter, d_c, influences the column pressure drop. Higher optimum inlet-to-outlet pressure ratios bring about considerably increased speeds of analysis. This can be realized in two ways: application of (wide-bore) columns at vacuum outlet, which is particularly advantageous in GC/MS, or by using narrow-bore columns.

For minimum plate height conditions, the analysis time, t_R, is proportional to $(d_c)^x$, with $1 \leq x \leq 2$, depending on the pressure gradient.

Variation of the inner column diameter and the stationary phase film thickness also largely affects the column working range, as well as the minimum detectable amounts and concentrations.

The sample capacity, ψ_s, i.e. the maximum amount of a component that can be injected on a column without causing appreciable (e.g. 10%) peak broadening, is approximately proportional to $(d_c)^3$ for constant capacity factors.

The minimum detectable quantity, ψ_0, of a given component is proportional to $(d_c)^z$. The exponent z varies between 1 and e depending on the ratio of column inlet and outlet pressures, and on the application of either a mass or concentration sensitive detector. Increase of the film thickness, except in a few situations, increases ψ_0. Also the smallest minimum detectable quantity is obtained for the smallest injection band widths. In general, narrow-bore columns are suited best if low detection limits are demanded. These capillary columns have the additional advantage of reduced analysis times, thus decreasing instrumental costs in GC/MS.

The working range $W = \psi_s/\psi_0$ of a column should exceed the concentration ratio of the compounds present in the sample to be analyzed. For high-pressure-drop columns, W is proportional to $(d_c)^2$, indicating that wide-bore columns are preferable in this respect.

The conclusion that narrow-bore columns should be used if detectable quantities are to be minimized does not imply that these columns are best for trace analysis. An important factor is determining the lowest concentration C_0, in the original sample that can be analysed by a given column-detector system, is the sample volume allowable to the column being largely in favour for wide bore columns.

Both theory and experiment show that under constant resolution conditions, the minimum value of C_0 for thin film columns is proportional to $1/d_c$ for mass detectors. For concentration detectors C_0 is independent of column diameter and film thickness. For both classes of detectors C_0 minimizes for input band widths between 0.1 and 1 of the chromatographic peak width. The effect of an increased film thickness yields only a marginal reduction of C_0 for mass detectors. If the concentration of a trace compound is below C_0 for a given column-detector system, sample enrichment or on-column focussing has to be carried out. Practical examples supporting the given theory will be presented.

Finally a comparison will be made between high temperature GC and SFC for the analysis of high molecular weight hydrocarbons.

Characterization of Heavy Oil Residues by Multi-Element Evolved Gas Analysis

A.J. Meruma, M.C. van Grondelle, H.C.E. van Leuven and L.L. de Vos
Koninklijke/Shell-Laboratorium, Amsterdam, (Shell Research BV), Badhuisweg 3, 1031 CM Amsterdam, The Netherlands

A growing interest in the conversion of biomass, shale and heavy oil into lighter and whiter products has stimulated the analytical characterization of these feedstocks. Prime parameters in this characterization are: the elemental composition, hydrocarbon type analysis (density) and molecular mass distribution (volatility). Classical analytical techniques, such as gas chromatography, UV spectrometry etc, measure only one parameter at a time in high-boiling and complex mixtures. Hyphenated techniques, i.e. a combination of two or more analytical techniques, allow the gathering of two-(or more) dimensional data from only one analysis. Well-known hyphenated techniques are: GC-MS and GC-IR. Recent additions with potential for use with, inter alia, coal, shale and oil products are combinations of thermal analysis with mass spectrometry (MS) or with Fourier transform infrared spectrometry.

In 1980 we developed a multi-purpose analytical method, which comprised the following steps: programmed heating of a few mg of sample in controlled atmosphere (usually helium), combustion of the gases evolved with oxygen, at 1000°C and measurement of the combustion gases with a quadrupole mass spectrometer. With this analyser the following parameters are determined:

- volatility of heavy oil products with concurrent determination of the elemental composition over the entire "boiling range",

- moisture content and "proximate" and "ultimate" analyses of coal,

- elemental composition of gases evolved, during programmed heating in a controlled atmosphere: "moisture" and "volatiles" evolved in an inert gas and "fixed" components evolved during ashing in oxygen,

- coking propensity of heavy oil products,

- ash content.

The instrument has become an indispensable tool in the screening of materials, in compositional analysis and in kinetic studies. The design of the instrument and applications of the techniques to the analyses of polymers, coal and catalysts have been described elsewhere. Examples of applications of the technique to the analysis of petrochemical feedstocks will be given.

1. H.C.E. van Leuven et al., "A Multi-element evolved Gas Analysis Technique. Design, Performance and Applications", paper presented at the Symposium on Compositional Analysis by Thermogravimetry, 16-17 March 1987, Philadelphia, Pennsylvania, U.S.A.

Development of an HPLC Method for the Determination of Nitrogen Containing Corrosion Inhibitors in a Mixed Hydrocarbon/Glycol Matrix

E.H. McKerrell and A. Lynes
Shell Research Centre, Thornton Research Centre, P.O. Box 1, Chester CH1 3SH

This paper describes the background requirements/restraints, development and auotmation of an HPLC procedure for the quantitative determination of two nitrogen containing corrosion inhibitors.

These corrosion inhibitors are used to protect oil pipelines from corrosion and are complex nitrogenous mixtures known to contain amides, amines and, in the case of one of them, imidazolines. Individual reference standards are not available with the result that formulated mixtures had to be used as reference materials in the method development work and subsequent routine analysis.

There were two major requirements:

1. to develop a rapid and sensitive method for the determination of both corrosion inhibitors in the presence of each other in fluids collected from the pipeline with the minimum of sample handling.

2. the development method should be a laboratory procedure which can be readily used in plant laboratories and hence should incorporate automation where possible.

A derivatisation procedure using fluorescamine, a non-fluorescent compound which reacts selectively with primary amines forming highly fluorescent derivatives, was investigated. It was found that a complex mixture of fluorogenic species was formed and part of this could be used to characterise the corrosion inhibitor being used and also quantify its content in the samples. It was found that the imidazoline based mixtures, although not containing primary amine function, also reacted with fluorescamine forming fluorogenic species of use for the characterisation and content determination in samples. The chemistry of the imidazoline/fluorescamine reaction has not been elucidated.

The derivatives were chromatographed on a 250 mm x 4.6 mm ID reversed phase HPLC column (packed with Spherisorb S.5.ODS2) using water/methanol/dichloroethane (5/33/2 v/v) containing 0.1% v/v acetic acid as mobile phase at a volumetric flow rate of 2.0 ml/min. Detection was by means of a fluoescence detector using an excitation wavelength of 390 nm and an emission wavelength of 376 nm.

The chromatographic step of the developed method was amenable to a large degree of automation and this included automatic injection, data capture, wash cycle/re-equilibration and shut down of equipment at the end of the analysis.

Using this method the level of detection was found to be 0.02%w for the imidazoline based inhibitor and 0.1%w for the other one. This analytical procedure has been successfully commissioned and is routinely used in a plant laboratory.

Index

Accuracy 9
Adsorbents
    Tenax TA 37
Air analysis 39
Alkanolamines 229
Alkylphenol propoxylate 102
Alkylsalicylates 244
Aluminium 198
Aluminon 198
Anions
    ion chromatography 226
Analogue instruments 73
Antifreeze 148
Aqueous discharge 109
    GC/MS FABMS 111
Aroclors 257
Aromatics
    mono-cyclic 217
    di-cyclic 217
    tri-cyclic 217
    poly-cyclic 218
Ashless dispersants 249
Ashing 190
Asphaltenes 70
Atmospheric dirt
    engine wear 149
Atomic absorption spectrometry 84, 144
    cold cell 84
Atomizers
    ICP-OES 192
Autoanalysers 228, 225
Autosamplers 232
Automation 11, 14, 71

Barium
    discharge water 92
    sulphate scale 7, 95
Base number-total 149
Base oils 243
    volatility 180
Biogenesis 4
Bitumen
    ^{13}C NMR 33
Borate fluxes, see ICP-OES
Boron
    discharge waters 90
    in borates 149
Brandes method-aromatics 212
Brine 230
Bromide 236

Carbonate scale 5
Carbon number distribution
    waxes 163
Carbon type analysis
    ^{13}C NMR 33
Catalysts
    rhenium content 66
Cetane number
    ^{13}C NMR 35
Chemometrics 261
Chlorinated biphenyls
    see Polychlorinated
    biphenyls
Chromatography
    gas liquid
        capillary 50, 136, 167, 259, 281
        data manipulation 217
        electron capture 256
        fingerprints 135
        geochemistry 70
            crude light ends 116
        headspace analysis 37
        high resolution MS 50
        mass spectrometry (MS) 19
        hydrocarbon waxes 159
        integration 162
        polychlorinated biphenyl 253
        Porapak columns 117
        refinery streams 135
        septum 115
        stationary phases
            degradation 160
        syringe 114
    gel permeation 4, 169
    high performance liquid
        corrosion inhibitors 285
        RI detectors 213
        UV detectors 213
    Ion 221
        anions 226
        autoanalysers 228
        detector types 225
        exclusion 226
        pair 226
    super-critical fluid
        waxes 169
Coke production
    ^{13}C NMR 34
Colorimetric analysis 190, 198
Condition monitoring 140
Containers, polyethylene 87
Computers, Commodore 8032 136, 211
Conductivity detectors
    ion chromatography 224
Coolant leaks 148

Copper in discharge waters 95
Cores, drilling 72
Coronene, mass spectrum 26
Corrosion 5, 7
    inhibitors 148
        nitrogen containing 285
    crankcase 149
Crown ethers 7
Crude oils
    Alwyn 35
    fingerprinting sulphur 45
    gas analysis by GC 38
    light ends 113
    Iranian heavy 54
    Iranian light 52
    Middle East 52
    North Sea 48, 77

Detergents
    neutral and overbased 239, 244
Dialysis 263
Dibenzothiophene
    mass spectrum 27
    fingerprint 46
Didodecyl disulphide
    mass spectrum 27 – 28
Diesel fuels 173
Differential scanning calorimetry,
    see Thermal analysis
Digital instruments 74
Dispersancy 147
Dispersants 239
    ashless 250
Drilling fluids 125
    atmospheres 129
Drilling mud system
    atmospheres 133

Ecosystem-decontaminants 254
Eicosane 177
Electron microscopy
    freeze fracture replication 239
Elemental analysis
    by ICP/OES 81, 82, 198
    by ion chromatography 236
    by mass spectrometry 283
    by optical emission
        spectrometry 141
Ethylene glycol 148
Expert systems 280

Florisil 258
Fluorescamine 285

Fluorescent indicator analysis
    for aromatics 212
Formation waters 230
Formulated oils 263
Fuel dilution 147
Fuel oil
    naphthenic acids 110

Gas chromatography
    see Chromatography
Gas oils 173
Gasoline
    combustion products 221
    spillage 41 – 42
Geochemistry 3, 69
Ground water analysis 42

Heavy oil residues 283
Hydrocarbons
    heavy $C_{20} - C_{100}$ 211
    light $C_1 - C_7$ by GC 119
    middle $C_8 - C_{16}$ 125
        chromatogram 128
    middle/heavy 211
    pattern recognition 137
Hydrogen sulphide 7
Hydroprocessed oils 266

Inductively coupled plasma/OES
    borate fluxes 191
    fuel analysis 189
    ICP 81, 82, 144
    ICP/MS 59, 63
    instrumentation 202
        detection limits 201
        element wavelengths 202
    xylene fuel 190
Imidazolines 285
Insulating oil 253
Infra-red spectroscopy
    for aromatics 212
Ion chromatography
    see Chromatography
Ion exchange resins 222
Ion suppression 222
ISO SC Code 153
Isotope dilution analysis 64
Isotopes, ICP/MS 64

Kerosine 125

Laboratory data management
    systems 13

Index

Laboratory information management
   system (LIMS) 70
Lubricating oil
   additives 239
   automotive volatility 179
   thermogravimetry 179

Magnesium in detergents 246
Mass spectrometry 236
   chemical ionization 5
   discharge thermospray 19
      solvents 22
   fast atom bombardment (FABMS) 99
   GC 19
   high resolution 4
   high voltage 34
   hydrocarbon cuts 34
   ICP 59
   ICP/MS detection limits 62
      suppression effects 67
   quadrupole 50, 60, 283
   spark source 66
   thermal desorption/GC 37
Metalloids by AAS 84
Microprocessors 73
Middle distillate fuels 173
Multielement analysis 199
Multitasking 13

Naphthalenes
   mass spectrometry 49
Naphthenic acids, FABMS 106–110
Nitrate 221
Nitrite 221
Noack test 186
Nonyl phenol ethoxylate 100
Nuclear magnetic resonance 47
   (NMR) ^{13}C 33, 261, 27

Oil condition index 147
Oil spills 8, 177
Optical emission spectrometry 144
   ICP/OES detection limits 62, 81
Oxalate 232
Oxygen bomb combustion 233
Oxygen flask combustion 233
Oxy-hydrogen combustion 233

Particle counting 152
Perchloric acid 152
Phenols, tetra-*t*-butyl
   dihydroxy diphenylmethane
      (AN-2) 26

FABMS 109
Phosphorus, ICP-OES 85
   discharge waters 93
Phytane 45, 137
Pipelines, oil
   protection 285
Platform water 87
PIllution 8
Poly-alpha-olefins (PAO) 261
Polychlorinated biphenyls 253
Polycyclic aromatics 218
Polyelectrolytes 7
Poly-waxes 166
Potentiometric titration 149
Pour point 244
Pristane 45, 137

Quadrupole mass spectrometer
   see Mass spectrometer

Reprogrammable devices 13
Resins 70
Robotics 11
Routine analysis 12

Scale inhibitors 6
Sediments 70
Silicates 148
Silicon 190, 198
Simulated distillation 159
Sniffing 3
Sodium 197
Sodium alkylbenzene sulphonates
   FABMS 105
Spectrometric oil analysis 144
   inter-element effects 145
Spillage 8, 77
Standardization 13
Sterane 46
Sulphate 231
Sulphate scale 5
Sulphonation for aromatics 212
Surfactants 100
   alkanolamides 102–103
   FABMS 101

Tenax TA 37
Tenax GC 125
Thermal analysis 173
   differential scanning calorimetry 173
   differential TA 173
   pan designs 185
Thermal desorption

Thermal desorption (*cont.*)
   GC/MS 37
   GC 126
Toxic elements 79 – 80
Trace analysis 3
   fuel oils by ICP-OES 189
Training 10
Trisnormoretane 47
Triterpane 46

Ultraviolet spectroscopy
   for aromatics 213
UV detection-ion chromatography 224
Used oils
   spectrometric analysis 139

Vanadium, ICP-OES 67, 198
Viscosity 146

Water-oily discharge 44, 77
Wax 245
   Fischer–Tropsch 170
   microcrystalline 168
   paraffin 167
Wax appearance temperature 173

X-ray diffraction 279

Zeolites 279
Zinc
   dipropyldithiophosphate
      mass spectrometry 24, 25
   in discharge waters 90